T0213406

ORIGIN OF THE HUMAN SPECIES

VIBS

Volume 106

Robert Ginsberg
Executive Editor

Associate Editors

A volume in
**Studies in the History of Western Philosophy
SHWP**
Peter A. Redpath, Editor

ORIGIN OF THE HUMAN SPECIES

Dennis Bonnette

Amsterdam - Atlanta, GA 2001

The paper on which this book is printed meets the requirements of "ISO 9706:1994, Information and documentation - Paper for documents - Requirements for permanence".

ISBN: 90-420-1374-5
©Editions Rodopi B.V., Amsterdam - Atlanta, GA 2001
Printed in The Netherlands

DEDICATED

to

My wife Lois
and our descendants

CONTENTS

FOREWORD

Michael J. Behe, Phillip E. Johnson, and others have led quite an exciting campaign against evolution's scientific merits, especially as a theory of how life first arose. Dennis Bonnette's philosophical book would have had the quick advantages of relevance, if he had designed it to contribute primarily to their campaign. Bonnette has taken a larger view.

The threat of evolution's implications in other branches of human knowing, not its purely scientific merits, has kept evolution controversial for a century and a half. Take its theological implications. How dire are they?

Darwinism has given scandal throughout the English-speaking world because many people see it to conflict with a literal reading of *Genesis*. Many Christians have not seen how to take divine revelation seriously without adhering to a literal reading of that part (and every other part) of the canon. As a Christian philosopher, Bonnette approaches Darwinism from a Catholic perspective that has never taken all parts of the canon literally. Instead, his perspective uses (1) an apostolic tradition of exegesis and (2) a heritage of classical philosophy.

Bonnette invites us to notice that evolution has philosophical implications, too. To see how dire the philosophical ones are, he holds his philosophy to a norm of time-tested rational insight. Bonnette adheres to St. Thomas Aquinas's explication of Aristotle's *philosophia perennis*, with results that are unfailingly interesting and often surprising.

Aquinas often used the term *"perfectio"* to mean a "completive trait," a trait that served to complete something with respect to its generation, development to maturity, or quest for fulfillment. In the First Part of his *Summa Theologiae*, q. 4, a. 2, Aquinas maintains that any completive trait present in an effect must be found in its effective cause through (1) univocal causality, "like a man who begets a man" or (2) a higher, equivocal causality, "like the sun containing a likeness of the things generated through the sun's active power." A genetic mutation whose propagation would be highly beneficial in a new population would probably count as completive in Aquinas's vocabulary. Suppose an animal is born with such a mutation. When we are told by Aquinas that this animal must have had an efficient cause that had already that trait in act, in the same or a higher manner, does this mean (1) that this animal must have had an ancestor that already possessed that trait? Or (2) does it allow for the possibility that every ancestor has been lacking that trait so long as higher and more universal causal influences, like the sun, supply the difference, so that somewhere, in the whole set of simultaneously acting efficient causes of the birth of that animal, that completive trait or something more perfect must pre-exist?

Many good Christians, scandalized by Darwinism, have wanted their religion to offer the first answer. But this cannot have been the opinion of Aquinas, the most eminent philosopher-theologian of the Christian religion.

This first answer is inconsistent with his theory that solar heat can produce low forms of life, often called the theory of spontaneous generation. The second answer is therefore more likely to be the correct interpretation of him. Still, this fact's bearing upon the modern discussion of evolution is difficult to assess because the term "evolution" is deeply and dangerously ambiguous.

Consider the hypothesis (1) that things with the trait of being "alive" emerged in a high-energy state of some original soup of things not alive and (2) that, since then, each trait representing an evolutionary step forward has emerged in plants or animals none of whose ancestors had that trait. Does this count as an improved version of the spontaneous generation theory? If it does, equivocal causation still has a biochemical interpretation. If not, the claim that an empirically emergent higher form must have pre-existed at least as completely in its efficient cause would simply require that God's influence be included in the account of that cause. Since no empirical science handles that part of the account, the question turns to what "evolution" names. Is it (1) an empirical theory that omits the part about divine influence, or (2) a new metaphysic designed to exclude any such part?

Bonnette explores these alternatives, and much more. He does so (if I may pronounce my own judgment as a Catholic theologian) with flawlessly orthodox Scriptural interpretation. His philosophical work is bound to have high interest at least among those who share my perspective. Still more, I hope his work will intrigue and enlighten a wider audience of Christians, Jews, Muslims, and many others. For no one group contains all of us who want to learn how to handle the evolution debate with all the theological, philosophical, and scientific sophistication it deserves.

William H. Marshner, S.T.D.
Chairman, Theology Department
Christendom College
Front Royal, Virginia

PREFACE

Perhaps the best way to greet the reader of this work would be to offer a short explanation of the selection of its title, *Origin of the Human Species*. Wording evidently derived from Charles Darwin needs some justification beyond the value of imitating secular success. I retain Darwin's term, "origin," because of its deliberate ambiguity. While Darwin's use of the term is now rightly understood to mean origin by evolution, I wish to address the challenge that contemporary scientific creationists pose to contemporary, mainstream scientific evolutionists. Creationists propose an entirely different understanding of "origin" than do mainstream scientific evolutionists. As a philosopher, I wish to explore the exact epistemological status of these diverse theories of origin.

Retention of the term, "species," might appear anachronistic because the success of Darwin's thesis initially contradicts its original signification. The conception of gradually evolving populations of individuals, groups of organisms that endlessly blend into other related forms as time passes, has gradually replaced the traditional notion of biological species as utterly fixed.

I use the term, "species," in the present title to resurrect the question whether, especially with respect to human nature, we can still give the classical notion of formal fixity continued legitimacy. I intend my title to indicate revisitation of some crucial issues that have long dominated center stage of human concerns about our own origin and nature and the question whether anything truly permanent underlies the ever-changing phenomena of the experiential universe. This last issue, reflected in the present often-acrimonious clashes between creationists and evolutionists, touches on a fundamental divergence in philosophical and theological traditions. As Raymond J. Nogar says, apparently much of Christian philosophy stresses things' fixity, while evolutionary thought stresses their flux.[1] Understandably then, claimants on both sides feel that much is at issue.

Little doubt can exist that scientific evolutionists initially express their views in terms of biological science. Still, relevant evidence and judgments concerning evolution's validity and significance exist in many allied disciplines, including, but not limited to, biochemistry, archeology, paleontology, paleoanthropology, anatomy, and even astrophysics. No single specialist in the natural sciences is fully qualified to evaluate the data and draw conclusions from so many diverse specialties. This almost inescapable situation inevitably begets inherent limitations upon any natural scientist or other specialized speculator who attempts to synthesize a comprehensive overview or definitive judgment of this intrinsically interdisciplinary doctrine. Nor surprisingly, experts in one field sometimes underestimate the value of contributions made by other researchers in widely diverse areas, or, at times, even within their own academic specialty.

We live in an age and culture that tends to enthrone natural sciences as the only credible source of rational truth. To suggest that philosophical science should possess a regulatory role that transcends the value of any positive science tends to be scandalous to the modern mindset. I do not seek to subvert or replace legitimate functions of the various, specialized natural sciences. Still, I want to examine the theory of evolution and some of its implications from the more general perspective of philosophical science. Before I can start, I need to make a few observations that support philosophy's legitimacy as a true and regulatory science.

The present age of technological progress and scientific wonders ought not forget its origin. The ancient Greeks first sought rational understanding of the natural world. These efforts marked the beginnings of Western Philosophy, whose accelerated success developed from a logical rigor and systematic approach not seen in previous history.

At its pinnacle, Hellenic thought produced Aristotle's (384-322 B.C.) natural philosophy and metaphysics. These philosophical disciplines established the division and methods of the different natural sciences and of that science which transcends nature: divine science, metaphysics. While Aristotle (the Stagirite) did not know or employ the modern experimental method, he did engage in a careful empirical study of nature and thereby defined prototypes for disciplines ranging from biology, anatomy, and physics to psychology and astronomy. Aristotle established the foundations of the philosophical science of physics in which he explained the meaning of such fundamental concepts as nature, space, place, time, and motion. Contemporary positive sciences presuppose and explore the same arena by experimental means. Current natural science also presupposes the first principles of being, like non-contradiction, sufficient reason, and causality, among others. Every experimentalist and scientific speculator uses such principles. No scientist can work without them. Yet, no natural scientist investigates or explains the rational and ontological foundation of these principles. Only the philosopher, the metaphysician, can defend them. Whether or not the modern natural scientist appreciates metaphysics, the philosophical science of metaphysics is the solitary and sufficient defender of the intellectual underpinnings of the natural sciences. From this it follows that, if philosophy is not, in fact and truth, a logically prior and superior genuine science, then no positive or natural science is genuine.

Only philosophy, not the physical sciences, can demonstrate the epistemological foundations of natural science. Natural science operates in a world of epistemological realism. Every serious researcher of the mysteries of the natural universe would reject the idealist suggestion that physical entities, from subatomic quarks to trans-cosmic quasars, are only images in the brain or alterations of the psyche. Only classical philosophy can sustain the critical realism needed to rescue physical science from naïve epistemological theories.

And only philosophy can transcend the perinoetic (knowledge of sensible or common accidents by substitute signs) restriction to the world of phenomena that intrinsically limits natural science. Natural science cannot dianoetically (using essential properties or proper accidents) penetrate the essence of realities. Only philosophy can. The most eminent scientists are well aware of this inherent limitation in their method. Reflecting similar observations made by Albert Einstein, Sir Arthur Eddington tells us that science can say nothing about an atom's intrinsic nature. He maintains that, just as we apprehend everything else in the realm of physics, we grasp the atom as "a schedule of pointer readings."[2]

To those unfamiliar with realistic metaphysics, to pierce the veil of scientific phenomena and understand the essence of things themselves might be grandiose self-deception. To the classical metaphysician, being's first principles, naturally self-evident to the human intellect, readily lend themselves to a discursive process that penetrates beyond all appearances and into reality's substance.

In this work, I employ the classical metaphysician's habit of analysis, as a critical instrument of reason, well suited to judge the relative logical and epistemological merits of contending views. I do not attempt to supplant legitimate speculations of experimental science in its own domain. My concern is whether, and in what fashion, evolutionary theory is compatible with classical principles of being and change, which apply to things in themselves (substances) and reflect the human mind's natural metaphysics. My interest is in the many points of potential conflict and confluence that arise in the interface of these diverse perspectives. As my title indicates, I need to explore the origin of species, especially the human species.

I am a Christian philosopher. As such, after I finish this project as an essentially complete act of unaided natural reason, by means of the same unaided natural reason, I need to determine the extent to which evidence of evolutionary science is consistent with supernaturally revealed teachings about our origin. I plan to do this as a philosopher, not as a theologian.

ACKNOWLEDGMENTS

I thank Christendom Press for giving me permission to republish all the articles I have previously published in *Faith & Reason*. Chapter Five is an edited version of my article, "A Philosophical Critical Analysis of Recent Ape-Language Studies," *Faith & Reason*, 19: 2, 3 (Fall 1993), copyright © Christendom Educational Corporation 1993.

I thank Niagara University for granting me two sabbatical leaves since 1989, allowing in-depth research of this book's many speculative problems.

I am indebted to the late Robert Lincoln Smith II for giving me my initial love of natural science. I am grateful to the late Joseph L. Dieska for showing me evolution's profound interface with central philosophical and theological issues. I thank my colleague, Raphael T. Waters, for reading and critiquing an initial draft of my research some years ago. Dr. Waters has continually encouraged and assisted my pursuit of this project, especially by providing difficult-to-obtain manuscripts written by Fr. Austin M. Woodbury, S.M.

I express gratitude to Michael A. Cremo for numerous helpful suggestions, especially regarding Chapter Fourteen's paleoanthropology. I am also grateful to my colleague, biologist William H. Cliff, for much needed guidance in natural scientific matters. Dr. Cliff provided detailed comments and suggestions incorporated in multiple rewritings of Chapters Thirteen and Fourteen. These commentators provided many positive insights; defects are invariably my own.

I thank my sister, Mary Anne Smith, for continual encouragement and assistance in computer-related problems and maintaining backup files. I thank my daughter, Elizabeth, for scanning my 1989 sabbatical draft manuscript onto computer disks. I thank Sharon Senick for proofreading the formatted text. I am also grateful to Hugh J. F. McDonald for his extensive efforts in preparing the book's index.

I am especially indebted to VIBS Deputy Executive Editor Peter A. Redpath for encouraging me to publish this work in the VIBS Studies in the History of Western Philosophy special series. I am grateful to Dr. Redpath both for his careful editing of my work, thereby greatly perfecting the final product, and for his amazing patience with me.

Finally, I express loving appreciation to my wife Lois, who patiently tolerated the widow's lot, having her husband computer-wedded for lengthy periods during this book's several drafts and final editing.

One

DARWINIAN EVOLUTION
VERSUS SCIENTIFIC CREATIONISM

From its start, Western Philosophy has witnessed intellectual obsession with (1) natural species' apparent fixity and (2) the presumption that the existence of immutable essences, either in or before things, makes natural species fixed. Aristotle states that the Pythagoreans were the "first to take up mathematics."[1] They insisted that "number was the substance of all things."[2] Aristotle records that "regarding the question of essence, they began to make statements and definitions."[3] He then tells us that Socrates focused his attention on the definitions of ethical matters.[4] Plato responded that, because sensible beings constantly change, "the common definition could not be a definition of any sensible thing."[5] Hence arose Plato's notion of spiritual ideas or forms, in which he said sensible things participate, or which sensible things imitate.[6] Plato, faithful to the tradition of Parmenides, conceived these forms as eternal and unchangeable.

Aristotle, the architect of moderate intellectual realism, dissolved Plato's world of pure forms. He replaced it with a universe of mentally-derived concepts as the only actual expression of species. Aristotle insisted that individual realizations of these species (for example, horses, dogs, and trees) existed solely because they concretely received substantial forms in their matter. He maintained that the formal identity of these concretized universals served as the real foundation (*fundamentum in re*) for the abstracted universal. The evident fixity of living species served as the external evidence for the stability and universality of these forms.

Thus began a tradition of metaphysical essentialism that permeated subsequent Western history. From the early neo-Platonists and Aristotelians to the later scholastics, rationalists, and idealists, all leading metaphysicians have insisted on the existence of universal, permanent structures of reality: essences or forms. Prior to Darwin, nature's order and constitution apparently proclaimed an experiential basis for such structures.

Charles Darwin did not originate the notion of evolution. This idea extends at least to Anaximander in the sixth century B.C. Anaximander declared that vapors caused by the sun produced animals and that human beings arose from fish.[7] Most of the notions that Darwin advanced on behalf of evolution were old. His one novel proposal was false. R. F. Baum asserts that Darwin's predecessors had already determined all the main elements of his theory, including natural selection. They could find no foundation for evolution in them. Darwin's novelty lay in his claim that natural selection was a primary and unlimited power, not a secondary and pruning one. On this point, on which so much of his fame rests, he was completely wrong.[8]

Immanuel Velikovsky notes that Darwin's contribution is the claim that natural selection is evolution's mechanism but that, like sexual selection, natural selection does not explain species' origin. Velikovsky thinks that Darwin's insight is that natural selection weeds out the unfit.[9]

Whatever the case about Darwin's originality, Darwin advanced (1) belief in evolutionary doctrine and (2) the supposition of nearly-infinite, minute changes in living things that gradualistically produce the transformism that evolution requires.

"Transformism" means "across forms." It highlights our problem. If innumerable transitional intermediates link forms of living things, how can we meaningfully speak about species entailing a potentially infinite multitude of individuals that are (1) qualitatively identical to each other in essence and, simultaneously, (2) qualitatively distinct from members of other species? Apparently, the post-Darwinian mindset understands living species as simply names we give for mid-ranges of extended spectra of unique individuals endlessly mutating into new forms of life. In such a world, the traditional notion of specific essence becomes an intellectual convenience with no extra-mental foundation.

Vertebrate paleontologist George Gaylord Simpson summarizes evolutionary science's problem with the concept: "No precise and generally acceptable definition of a species has yet been achieved."[10] Simpson says this because evolution causes species to "intergrade in nature." He observes that species are not static. Sometimes they differ "obviously and radically." At other times only "subtly and dubiously." No sharp divisions exist between them, which we would expect if a separate creation had produced them.

Evidence favoring interconnectedness of all living things appears overwhelming. Many diverse sciences find it. Experimentation with selective breeding of plants and animals has produced new strains of grains and breeds of livestock. While most such new forms tend to blend back into former states unless deliberately segregated, a few, such as the distinct plant, *Raphanobrassica* (developed by crossing the radish with the cabbage), do not.

While Darwin could offer no mechanism for production of new species (transformism), in 1900, Hugo De Vries, E. Tschermak, and K. Correns rediscovered Gregor Mendel's work on heredity. When Mendel's work was combined with De Vries's discovery of mutations, useful explanations arose for emergence of new forms.

Examination of biogeographical and ecological data reveal that plants and animals exhibit characteristics that are clearly differentiated in correlation with their geographic distribution and history. The science of comparative anatomy has established innumerable instances of fundamental resemblances combined with lesser differences that suggest affinity and descent among all living species. Paleontology examines the fossil record and observes broad outlines of continual change and development throughout history as preserved in geological strata. The most famous, detailed example of such development is the emergence of the modern horse from its diminutive, presumed ancestor, *Eohippus*. Recent studies of the genetic code as embodied in the DNA macromolecule indicate ratios and sequences of nucleotides that bespeak submicroscopic analogies, and sometimes near identities, between all living organisms. The synergistic effect of this bio-chemical evidence, when combined with the others mentioned above has been so complete that we tend to dismiss any opposition to the general theory of evolution as pseudo-science.

Still, the 1970s witnessed a partially-successful attack against scientific evolutionism by scientific creationism. And while scientific orthodoxy dismissed this movement as pseudo-science, some scientific evolutionists felt the need to defend evolution against scientific creationism.

In 1981, creation science delivered a blow that rocked scientific academe when Arkansas approved a state law, Act 590, requiring balanced treatment for creation science and evolution science in public schools. Section 4 of Act 590 defines creation science to include, among other things, "changes only within fixed limits of originally created kinds of plants and animals" and "separate ancestry for man and apes."[11] If these creationist points could be sustained, consequent stability of species would enhance defense of the traditional notion of immutable biological essences or forms.

In support of their case, creationists frequently indicate that geological evidence shows that species apparently emerge suddenly and instantaneously whole, remain substantially unaltered for extended periods of time, and pass away instantaneously. In sharp contrast to the conventional Darwinian postulate of phyletic gradualism, other species, anatomically similar, and presumably related, appear in later geological formations more by addition, or possibly by substitution, than by transformism. In support of their position, creationists frequently point to the recent, "punctuated equilibrium" theory of Stephen Jay Gould and Niles Eldredge. According to creationists, punctuated equilibrium (1) recognizes species stability and (2) supports their view that transformism does not occur. Punctuated equilibrium claims that evolution "proceeds by rapid fits and starts, punctuating long periods of stasis."[12] Contrary to what evolutionists might generally be expected to say, Gould affirms sudden appearance of new species. Still, save for the fact that he argues for "a jerky, or episodic, rather than a smoothly gradual, pace of change," he styles himself an evolutionist. Gould claims that, if evolutionary theory is correct, sudden emergence of new species, followed by a period of stability or "stasis" is exactly what we should expect. To the creationists' chagrin, Gould passionately denies creationists' assertions that he admits the fossil record shows no transitional forms. While such transitions are usually

4 ORIGIN OF THE HUMAN SPECIES

absent at the level of species, he maintains, "they are abundant between larger groups."[13]

Gould and other contemporary evolutionists maintain that the apparent, nearly instantaneous emergence of new forms in the geological record actually consists of very rapid branching, occurring over a period of hundreds or even thousands of years, a time long by human life's measure, but so short as to be almost invisible in the span of geological eras. Hence, punctuationists use the terms "sudden or instantaneous" in a strictly relative sense.

Many contemporary natural scientists defend the existence of transitional forms between biological species and higher taxa (genera). Roger J. Cuffey concedes the critical role played by transitional fossils in sustaining the thesis of organic evolution. He indicates how paleontologists have been "appalled" at "grossly misinformed" claims, often made by otherwise well-informed people, that transitional fossils do not exist.[14]

Cuffey offers several tables of examples of what he terms transitional individuals, successive species, successive higher taxa, and isolated intermediates. The epistemological status of any groupings above the level of species is such that real proof of transformism ultimately relies upon concrete evidence of inter-specific change. Cuffey is well aware of the intramental character of transitions between genera. He concedes that higher taxa (those above the species level) do not exist in nature in the same way as do species. He admits that such more general classes of organisms are "simply the scientist's mental abstractions," grouped together based upon the "various degrees of over-all morphologic similarity displayed."[15]

Anti-evolutionists aim their darts precisely at the level of inter-specific transformism. To cite but one example, Cuffey lists a study of echinoids by A. W. Rowe in his table of examples of transitional individuals.[16] But A. N. Field maintains that the fossils of echinoids (sea urchins) exhibited intra-specific variations, not inter-specific change.[17]

Field argues that when the Micraster authority, Rowe, sorted through some 2,000 such fossils, arranging them according to geological age, he found that, "when all was said and done he was just as much a Micraster sea urchin as when he began." Rowe noted variations in minor anatomical features, such as the shape of the mouth or layout of the spines, but nothing to indicate that the last of the series was more than a variety of the first. Since no one doubts that variation within species occurs, the evidence would in no way support any claim of inter-specific evolution.[18]

According to Douglas Dewar, the same evidence occurs with the best known of all the examples of purported inter-specific transformism, the series from the dog-sized Eohippus to the modern horse.[19] Dewar concludes that, like the sea urchin series, the horse series starts and ends with a horse.[20]

The above two examples indicate a tendency among some scientific evolutionists to interpret as genuine transformism the same fossil data which anti-evolutionists read as simply variations within a species. This conflict highlights the importance of determining the proper biological and philosophical meaning of the concept of species.

If inter-specific transformism is crucial to evolutionary theory, evolutionists can take little solace in Gould's admission that transitional forms are usually absent at the level of species, even though he claims they are "abundant between larger groups."[21] At the same time, scientific creationists can take no comfort in Gould's apparent refusal to concede that transitional inter-specific forms do not exist.

Another point of conflict between evolutionists and creationists is the evident gaps that exist in the fossil record. Darwin had noted the "abrupt manner" in which whole groups of species appear suddenly in some geological formations. Louis Agassiz, F. J. Pictet, and Adam Sedgwick had indicated this fact as a "fatal objection" to transformism. Darwin granted that, if the claim that numerous species in the same genera or families actually started life at the same time were true, his theory of descent by slow modification through natural selection would be untenable.[22]

In this case, and others, Darwin and later evolutionists blame incomplete geological records. Sudden appearance of new populations, as postulated in recent theory by punctuationists and others, entails a parallel need to explain absence of intermediate, inter-specific forms. Scientific evolutionists explain absence of fossil evidence by appeal to the extreme difficulty of fossil formation in the relatively small population groups in which rapid speciation takes place. T. A. Goudge makes the typical case by pointing out that most plants and animals leave no record of their presence because they rarely meet the conditions needed for fossilization to occur.[23]

Anti-evolutionists respond by contradicting the explanation's premise. They maintain that, far from being difficult and rare, fossilization often occurs. Douglas Dewar and G. A. Levett-Yeats, Fellows of the British Zoological Society, investigated the fossilization question. They concluded that fossilization is so common that, where fossil-hunting had been most intense, namely, Europe, the genera of land mammals (excluding bats and aquatic ones) had 100 percent representations in fossils.[24]

Dewar challenged in principle the possibility of standard evolutionary theory achieving all the intermediate stages needed to explain its postulated scenario of emergences. He defies anyone to explain the detailed steps needed to accomplish the conversion of a land mammal into a whale, an evolutionary change which all evolutionists claim took place. He describes the complicated modifications required for the young to go from being suckled in air to being suckled under water. He insists that no conceivable intermediate stages could exist allowing for a gradual transition. One example among many he gives is the need for the young to "have its windpipe prolonged above its gullet to prevent the milk ejected by the mother from entering its lungs." All such modifications need to exist prior to any young being born under water.[25] Dewar views such changes as sudden and miraculous, or, as produced by some "equally miraculous prophetic evolution" with everything somehow designed and executed in advance of its actual deployment.

Although not a scientific creationist, biochemist Michael J. Behe, in his *Darwin's Black Box*, stirs recent controversy by moving Dewar's argument into the Lilliputian world of the cell. He claims that complex biochemical systems exist which defy explanation by chance or gradual changes.[26] Such biological systems as bacterial flagella, bloodclotting, cellular transport, the immune system, AMP/ATP molecular synthesis, and so forth, manifest "irreducible complexity." This complexity entails mutual interaction and interdependence of all functioning parts such that, should any part be missing, the whole cannot function. Such systems render gradual evolution unimaginable since a stepwise progression to completion seems impossible. An Intelligent Designer appears needed.

Evolutionists counter that complex systems evolve from simple ones, that parts change as systems develop. They maintain that completed systems may not resemble initial parts. Parts may serve different functions at various stages. Subsystems may be helpful, but not necessary, at one point. Later the same element becomes essential. The "scaffold" argument suggests that genetic steps leading to "irreducible" complexity were initially present, but later disappear – leaving the false appearance of irreducibility.

Such alternative scenarios are easier to hypothesize than demonstrate. And Behe concedes that "one cannot definitively rule out the possibility of an indirect, circuitous route."[27] But he argues that, "as the number of unexplained, irreducibly complex biological systems increases," our confidence in Darwinism collapses.[28] Behe's argument is not simply a "God of the Gaps" fallacy. Mutational steps requiring too great a complexity become mathematically incredible. Thus, irreducibly complex systems that permeate our world argue forcefully for an Intelligent Designer.

Such counterpoints as Dewar's and Behe's might cause serious students of science to lament the manner in which religious fundamentalism spawned scientific creationism, thereby weakening from the start its credentials as legitimate natural science. Witness the harsh, but not entirely undeserved, conclusory judgment of biologist Joel L. Cracraft. It reflects a common attitude of many natural scientists. Cracraft describes scientific creationism as a "charade," "political propaganda of a conservative, fundamentalist belief system."[29] Nonetheless, a PBS-TV documentary entitled, "God, Darwin and Dinosaurs," noted that some 30 percent of secondary biology teachers prefer creationism to evolution.

From the start of Darwinian evolution, had solely rational objections to certain theories, tenets, and claims about evolutionary theory been advanced, more fruitful dialogue might have ensued. In fairness to the proponents of scientific creationism, they have advanced fundamentalist tenets in conjunction with a total positive cosmology precisely to avoid a fideistic skepticism that knows no scientific rationale. Even if some researchers have questioned the authenticity of their science, most adherents to scientific creationism sincerely believe in their project's legitimacy. Not all errors in scientific methodology convert the efforts of scientific creationism into an act of faith. Sometimes, bad science accounts for their views, the practice of which is not the sole domain of the scientific creationist.

I cannot try to settle definitively all the natural scientific questions that relate to the validity of evolutionary theory. And to do so is not germane to my present study. As a philosopher, I can competently examine and compare conflicting claims of natural scientists with respect to their logical, epistemological, and metaphysical foundations. I will try to do this representatively, not exhaustively, to learn what, philosophically, we can safely say about evolutionary theory and its impact upon the biological and philosophical notion of species.

Evolutionary theory's epistemological foundation is not something totally extra-mental. Fossil evidence, for example, is comprised of nearly innumerable individual and unique remains, impressions, or traces of animals or plants. The process of classifying or grouping these remains into species, or the act of perceiving a series of them, as blending from one population into another, is a mental, not an extra-mental, act. The process of classification attempts to understand unity within a plurality of individual beings, thereby imposing an interpretation upon the experiential data. An interpretation is an intelligible structure or process formulated in the intellect which, we hope, faithfully describes what exists extra-mentally.

Raymond J. Nogar makes clear the case for some type of general evolutionary doctrine. He argues that sound science, philosophy, and theology are reluctant to concede that hundreds of thousands of special acts of creation would be needed in order to account for the origin of all the species which have ever existed. Nogar maintains they do this because of the principle of economy. He concludes that some form of evolution is the only acceptable alternative explanation of the fossil record.[30]

Nogar conceives that evolution of all things in the universe, both non-living and living, came about in a "one-way, irreversible historical process, by a successive unfolding of the simplest units."[31] Thus, he advances the notion of evolution beyond mere biology to include "the scientific extension to all sectors of the phenomenal universe."[32] Nogar presumes that evolution for living things takes place by procreation and heredity. His cosmological extension of evolution is not of direct concern to us.

Most scientists today would say that evolution is more than a good hypothesis or valid theory. They would insist it is a fact established by science.[33] Nogar supports this claim with voluminous papers and discussions by leading scientific experts published during the Darwin Centennial Celebration in Chicago in 1959.[34] Again, Nogar insists that "the fundamental fact of evolution seems to be settled once and for all."[35]

Nogar points out the key role of the science of paleontology in establishing this "fundamental fact." Only paleontology, the science which reads the fossil record of the past, can provide the primary and direct evidence for evolution. Only paleontology can establish whether the evidence for evolution "on the grand scale" has taken place – or not.[36]

Nogar is well aware of the logical limitations of paleontological evidence. He knows the problems involved in trying to determine exactly what occurred millions of years ago. He admits the difficulties of "making inferences from analogies, extrapolation and other weak logical links."[37] As a result, he reverses his own field somewhat by concluding from all this that paleontology cannot alone prove that evolution has occurred.

Earlier Nogar admitted that direct evidence for evolution can be provided only by paleontology. But he is also well aware that "the earliest geological period of fossils, the Cambrian, shows most of the modern invertebrate phyla present in great detail and numbers."[38] While some allegations exist against the very existence of Pre-Cambrian fossils, no one suggests they were numerous. Hence, to say that the fossil record is incomplete with respect to these invertebrates is evident understatement. Out of an estimated 1,000,000 total species, invertebrates constitute 965,000.[39]

The paleological problem of the invertebrates is substantive. Each phylum appears structured following a basic "blueprint" dissimilar to any other phylum. Also, while within any given phylum, "good continuity" may be exhibited, yet, the family relationships between the different phyla are obscure.[40]

And yet, in maintaining that evolution is an "established scientific fact," Nogar does not claim that such a "fact" possesses the objective certitude proper to metaphysics or sacred theology. Instead, he explains that what he means by "fact" is simply the removal of serious doubts. He does not demand "absolute certitude" in every instance. Nogar views evolutionary fact as a "circumstantial fact," the product of logical inferences not unlike the judgments made in a court of law, especially like those made in criminal cases.[41] He seeks the gradual accumulation of evidence which persuades "*beyond a reasonable doubt*" and which concludes on the basis of "*a strong preponderance of converging evidence.*"[42]

Nogar concludes that the fact that all living organisms are the product of a general process of genetic descent with modification can be held with a "minimum of reasonable doubt" as a result of the "broad, converging argumentation" extracted from nearly every subscience of biology leading toward the same conclusion.[43] Any single element of the argument, such as genetics or natural selection, cannot stand by itself. Yet, Nogar insists that a "convergent proof" leading up to the "fact" of evolution can be established when all the elements are taken in relation to each other and added to the argument offered by paleontology.[44]

Nogar anticipates the logic of Sidney W. Fox's response against the claims of recent scientific creationists. Fox complains that creationists tend to attack one element of the evidence for evolution, while they ignore surrounding supporting proofs. For example, they may attack the lack of transitional forms in the fossil record while, at the same time, they ignore evidence provided by all the other allied sciences, such as neurobiology, biogeography, comparative anatomy, and so forth.[45] Like Fox, Nogar insists that all the relevant sciences "converge on a single conclusion:" evolution is the singular best explanation.[46]

Nogar admits that only paleontology provides "direct evidence" for the fact of evolution. The most that all other allied sciences can provide is merely "accessory" information.[47] Nogar concedes that evolutionary science cannot be proposed with the certitude proper to metaphysics or sacred theology. To the logician, cumulative circumstantial evidence, however impressive, can never equate to certitude. I do not deny the probative force of the argument from converging evidence. But we should be aware that the limitations of these perinoetic speculations, especially as they extrapolate from present data to prehistorical conclusions, forever foreclose the possibility of apodictic certitude.

Limitations of space prevent full discussion of the pros and cons of each of the major arguments advanced in the cause of evolutionary theory. Yet, we should remember that weakness of arguments favoring evolution does not necessarily prove that evolution does not occur. Anti-evolutionists sometimes forget that. On the contrary, the consensus of contemporary judgment clearly favors some form of evolution. Nonetheless, powerful logical objections against the force of evolution's arguments have been posed by Australian philosopher and theologian, Austin M. Woodbury. These objections deserve our careful attention.

Woodbury argues against the efficacy of what he terms the "homological" argument, any argument based upon similarity of general anatomical structure between diverse types of organisms. By way of analogy, his objections apply to comparative anatomy and to the comparative sciences of physiology, serology, pathology, and biochemistry. Woodbury argues that the homological argument fails in its objective for many reasons: (1) because it is possible that God simply created organisms according to an "archetypical" plan whose unity in the ideal order would naturally give rise to morphological similarities; (2) since higher forms "virtually contain" lower ones, one should expect production of similar formal effects such as similar organization and appearance; (3) similarities of limbs and organs can be explained by the fact that organisms must function under the same conditions of the same physical world; and (4) because species homologies may lend themselves to contradictory interpretation of species ancestry.[48]

Against the embryological argument, based upon similarities of development of various embryos, Woodbury replies that: (1) the embryo of a higher animal is never, in fact, in the same state as that of a lower species; (2) any such seeming likenesses are caused, for the most part, simply by "incompleteness of development;" and (3) such similarities also arise largely because of the need to satisfy the temporary similarities of the needs of the embryonic state itself.[49]

Against the once commonly held, but now largely discarded, argument based upon the presence of vestigial organs, Woodbury argues that: (1) reduced organs do not always lose their function while those (if there be such) which lack function are not always reduced; (2) disused organs are not always reduced, citing the well-known example of lambs' tails being amputated for centuries with no subsequent shortening of their offspring's tails; (3) Darwin and his followers were simply wrong in claiming that many organs, such as the endocrine glands, the vermiform appendix, and the coccyx, were functionless; (4) the argument falsely supposes that acquired traits are inheritable; (5) such organs may, indeed, have had a function in the embryonic stage; and (6) such rudimentary character of certain organs may be the product of unfavorable mutations.[50]

Woodbury grants that the argument based on genetics and mutations may have merit, at least with respect to showing that "some evolution within the same species is probable."[51] He holds that natural species are actually at the level of biological classes (more on this later). Woodbury maintains that "chromosomal mutations" may account for changes by loss or with respect to quantity. Yet he denies that such mutations can account for "transpecific" changes.

Woodbury responds to the paleontological and biogeographical arguments by saying they offer evidence, but "in favor of evolution only within the same natural species."[52] He envisages a breadth of extension for "natural species" which far exceeds any definition given by contemporary biologists.

Nogar draws conclusions favorable to evolution from paleontological evidence. He cites the evolutionary sequence of the horse, which he calls one of the "most thoroughly explored life histories of any vertebrate."[53] Also, he cites the echinoid Micraster sequence, which he refers to as probably "the most conclusive evidence to date."[54] Dewar and Rowe respectively rejected these proofs on precisely the sort of grounds Woodbury raised: the evidence is "valuable in favor of evolution only within the same natural species."[55]

The current consensus of expert opinion insists that some sort of natural process of descent with modification remains the most probable explanation of all available present evidence, despite the above logical and scientific objections raised against evolutionary theory. Curiously, this conclusion appears also urged upon us by natural science and natural philosophy.

Nogar argues that the philosopher of nature seeks natural, not miraculous, causes for natural effects. He points out that even the theologian accepts the principle that God normally operates through secondary natural causes and natural law, not through "miraculous intervention."[56]

St. Thomas Aquinas makes the same point when he says that God governs inferior things through superior ones. God does this, not because He is lacking in His own power, but out of the very abundance of His goodness. For, by this means, He communicates the dignity of causality. God allows creatures to participate in the causality which constitutes His very nature as First Cause Uncaused.[57]

Nogar admits that the creationists' way of interpreting the fossil evidence could be, at least in principle, correct. He grants that "sequential creationism" is a conceivable way of interpreting the fossil record, since God is perfectly free to create in any way He wishes.[58] But Nogar flatly rejects this explanation on the grounds that the miraculous is unnecessary, since a natural way of explaining the appearance of new species is obtainable, namely, through descent with modification.[59] Nogar relies again on Ockham's Razor, the principle of God's natural economy of creation. God simply posits, at the very beginning of creation, a process which unfolds the panoply of all creatures in a purely natural and immanent manner. God need not make successive and miraculous interventions into the natural world every time a new species is required.

Sequential creationism also appears to be an implied denial of the fullness of divine providence. The creationist insistence on continual miraculous intervention by direct creation of new natural species actually denigrates from the perfection of the world taken as a natural whole. This, in turn, denigrates from the power and providence of God Himself. Etienne Gilson explains that creatures participate in the divine perfection in such a way that to detract from the perfection of God's creatures is actually to detract from the perfection of the God's power. Thus, he observes, "A universe without genuine causality, or with a causality not allowed its full effect, would be a universe unworthy of God."[60]

Thus, perhaps the most forceful argument in favor of evolutionary theory lies in its ability to express most perfectly the fullness of divine creative causality, not in perinoetic speculations drawn from empiriological evidence. The principle of natural economy follows necessarily from the perfection of God's production of the world as a whole. The scientific creationists' repeated calls for miraculous intervention in the created order may actually detract from God's power. Evolution may give greater glory to the First Cause by preserving the natural efficacy of secondary causes. St. Augustine's appeal to "seminal reasons" rightly points to a more perfect Creator by minimizing the need for subsequent miraculous interventions.

And yet, in the economy of God's creation, genuine miracles still can and do occur, for example, the Incarnation and Resurrection of Christ. Nor do I intend to foreclose the possibility that legitimate scientific evidence and argumentation may show that key stages in the origin and development of organic life require direct supernatural intervention by God. But, the omnipotence of God might best lie in God having providentially ordained a world so structured from its inception as to give rise through the natural interaction of creatures to ever more perfect forms of reality and life.

The philosopher ought not transcend the proper limits of philosophical method by rendering definitive judgment as to purely scientific issues in evolution theory. Still, an examination of the metaphysical foundations of some form of evolutionary explanation of the origin and development of living organisms is in order. Any such possible explanation must be described as an expression of what is termed "theistic" evolution. Such a process cannot escape the absolute metaphysical necessity that God continues to create and sustain the existence and natural operations of all the natural agents involved.

Attorney Phillip E. Johnson points out how most leading evolutionists reject any form of theistic evolution.[61] The majority of evolutionary biologists appear to be resolute atheists.[62] Those who claim religion and science to be compatible almost invariably subordinate religion to natural science in such fashion as to gut revelation's claims. The vast majority of these scientists deny any suggestion of direct supernatural intervention in cosmic or human events, or even the general supervision of the created order by some sort of divine providence.[63]

Johnson shows, with disturbing clarity, how naturalistic philosophical presuppositions of these evolutionists, starting with Charles Darwin, have conditioned the "scientific" arguments favoring evolution. Such thinkers presuppose that no preternatural or supernatural forces exist to effect or influence life's origin and development. These evolutionists conclude that some sort of purely natural mechanism must tie together apparently similar and, thereby, related species of living things in some evolutionary panorama, whether adequate and definitive scientific documentation can be found or not.

The Darwinians' philosophical perspective excludes even the possibility of divine influence. Johnson observes that these evolutionists accept the philosophy of naturalism which holds that no supernatural being "could in any way influence natural events, such as evolution, or communicate with natural creatures like ourselves."[64] The whole of material nature is conceived as a closed system, utterly unable to be influenced by any external force.

Some form of descent with modification appears self-evident from this implicitly atheistic perspective. This virtually deistic concept of creation as standing alone has much the same consequences for those evolutionists who style themselves as theists, yet, who accept the central tenets of naturalism.

Atheistic evolutionists see the need for some evolutionary process because no God exists who can create new species. Theistic evolutionists deem evolution necessary as the most perfect manifestation of divine providence and omnipotence. Johnson will challenge the factual inferences of both these opposing groups.

Nogar, Fox, and others claim that an abundance of complementary data drawn from other allied sciences make up for any logical weakness in evolutionary theory caused by flaws in the fossil record. Johnson attacks this cumulative argument by (1) pointing out the naturalistic presumption that tends to be at the heart of all readings of the fossil evidence and (2) attacking an illicit inference from micro-evolutionary evidence to macro-evolutionary conclusions.

The difficulty with the fossils lies (1) in the questionable presence of transitional forms needed to prove the central thesis of Darwinism and (2) in the converse fact that we may read the evident stability of species over millions of years as *prima facie* disproof of gradualism's claims. Johnson notes Gould's admission that stasis (the tendency of species to remain relatively unchanged during their existence) and the sudden appearance of "fully formed" new species are two characteristics of the fossil record quite inconsistent with gradualism.[65] Johnson correctly observes that stasis "is also the norm and not the exception."[66] This norm of species stability is clearly foreign to the general belief that evolution requires gradual change through small mutations over long time periods.

Evolutionists such as Cuffey insist upon the existence of many "transitional fossils." Anti-evolutionists such as Rowe interpret most of the examples given as instances of variation within species. The evidence for transitions at the macro-evolutionary level is far more sparse. Johnson suggests only the therapsids (mammal-like reptiles often cited as links between reptiles and mammals) and the famed *Archaeopteryx* (cited as a link between reptiles and birds) as possible candidates for fossil evidence of macro-evolution. Neither example constitutes undisputed proof of such transitions.[67]

Eldredge and Gould's "punctuated equilibrium" theory is an attempt to sustain Darwin's general thesis despite the embarrassing character of the fossil record. They claim that rapid speciation takes place among a small group of individuals within the general population referred to as "peripheral isolates."[68] The relative absence of transitional fossils can then be explained by noting that (1) fossilization requires special conditions and (2) the likelihood of such conditions obtaining among peripheral isolates is so remote that we should expect to find such transitional fossils rarely, if at all.[69]

Eldredge and Gould's position suspiciously resembles a criminal defendant who presents an unproven alibi with a plausible, yet also unproven, explanation for the failure to supply initial proof of the alibi. We can imagine how well such unsupported claims would be received in a court of law! If the presumption of innocence were absolute, such unsubstantiated claims might well be accepted, since an absolute presumption would rightly tend to discount all *prima facie* evidence to the contrary.

Similarly, the Darwinians' naturalistic presumption in favor of evolution makes their interpretation become self-evident, necessary, and essentially unfalsifiable. Evolution is sustainable in spite of the more obvious reading of the fossil record, since every contrary interpretation can and must be explained away by some alternative explanation, such as "punctuated equilibrium."

The naturalistic presumption lies at the heart of the evolutionist's absolute faith that we can discover no apparently contrary evidence that can undo evolution's basic claims. This conviction that no logical obstacle can overturn the basic truth of Darwinism has led to the just accusation that evolution is no more a true science than is the much criticized creation science. Evolutionary theory and creation science fail philosopher Karl Popper's criterion of a true science: a genuine science must in principle be capable of being shown to be false through empirical testing.[70]

Johnson maintains that any fair reading of the fossil record stands in stark contrast to the natural expectations of Darwinism. Even Darwin recognized this fact. But his enthusiastic disciples do not. Johnson also attacks the argumentation supporting evolutionary theory which claims that other allied sciences, such as microbiology, biochemistry, embryology, and so forth, complement and support the data found in the fossil record so as to constitute "cumulative probative force." Woodbury's counter-arguments have already rebutted some of these data and cumulative arguments.

Johnson first offers several instances of the fallacious logic. Evolution's proponents use ambiguity in the term "evolution" so as to shift attention from legitimate instances of micro-evolution to illegitimate claims for macro-evolution. The trick, according to Johnson, is to take one of the more modest meanings of the term, "evolution," giving an example of its application which is generally known and accepted. Then, evolutionists claim that that limited example is evidence for the broadest acceptance of evolutionary theory "in which matter evolved to its present state of organized complexity without any participation by a Creator."[71]

Most everyone knows that bacteria will evolve into forms resistant to various antibiotics, that English peppered moths will adapt to changing industrial conditions, that mutations can occur in genetic material and result in minor changes in morphology, and so forth. Thousands of generations of fruit flies deliberately irradiated by H. J. Muller to induce mutations produced many monsters with varied coloring of eyes and varied lengths of wings and other minor changes, including the total absence of wings. Still, the final generation remained just as clearly a fruit fly as the first!

No one has ever witnessed a living species giving rise to a clearly diverse species. The inability to breed with forebears is not proof of an essentially new species, except in terms of the accident-based, arbitrary definition of many biologists. Evidence for some form of intra-specific evolution does not constitute proof for inter-specific evolution.

Johnson also attacks the role of the allied sciences (such as biochemistry, molecular biology, embryology, and comparative anatomy) that are supposed to help create the cumulative argument for evolution. They often do no more than document the inconclusive fact that similarities and relationships exist between certain species. This fact does not prove an evolutionary relationship or the existence of common ancestors, as the evolutionists invariably assume. We have seen earlier how Woodbury answers the argument from such similarities, with what he calls "the homological argument."

Darwinists assume that "relationship" amounts to the same thing as "common ancestry."[72] But Johnson exposes the logical fallacy involved when Darwinists refer to the evident "relationship" between humans and apes. We can observe that humans and apes are physically and biochemically similar. But we cannot observe the hypothesized "ape-like common ancestor." To the philosophical materialist, the existence of such a hominid is plausible. But the connection of such a hominid to present day apes and human beings, even if a candidate be found among fossil remains, is not directly observable. Johnson suggests that the "true explanation for natural relationships may be something much more mysterious."[73]

These same similarities, with which earlier biological essentialists were well familiar, were attributed to some sort of metaphysical blueprint which they termed the "Archetype," conceived as existing in, possibly, God's mind. They need not be the result of biological inheritance.[74]

The arguments of the allied sciences taken from similarities and from micro-evolution fail to offer logical support for the thesis of common ancestry. The "cumulative argument" of Nogar and others loses much of its probative force. The addition of fallacy to fallacy does not sum up to certitude nor even to increasing probability.

Johnson has most skillfully accomplished unmasking the illogical and atheistic presumptions at work in the heart of evolutionary theory. He restates fundamental problem areas, such as (1) the seeming absence of transitionals to fill the gaps in the fossil record, (2) the apparently sudden appearance of many complex types of organisms at the beginning of the Cambrian age, (3) the problems associated with attempts to explain the genetic code's origin, (4) the controversy over the recent claims about punctuated equilibrium, (5) the clearly defined limits which appear to restrict changes accomplished through breeding experiments, and (6) the significance of catastrophic extinctions on evolution theory.[75]

Crucial to Darwinian claims is that relationships between various living things were the product of some gradualistic transformism through descent with modification that occurred without God or any other "non-naturalistic mechanism" intervening. If that did not happen, Johnson insists, "we do not have any important scientific information about how life arrived at its present complexity and diversity."[76]

Johnson is not the first, nor will he be the last, thinker to raise substantive questions concerning Darwinism. Consider Australian molecular biologist Michael Denton. Denton reminds us that "the great biologists and naturalists of the late eighteenth and early nineteenth centuries who founded the modern disciplines of comparative anatomy, taxonomy and paleontology adhered strictly to a discontinuous typological model of nature."[77] Contrary to evolutionists' allegations, these scientists, including Georges Cuvier, Louis Agassiz, Charles Lyell, and Richard Owen, embraced typology because the fossil record appears to demand it, not from religious or metaphysical presuppositions.[78]

Denton maintains that Eldridge and Gould's punctuated equilibrium appears unlikely "to explain the larger systematic gaps" in the fossil record.[79] He allows that evolution could still be justified by "finding the connecting links, or by reconstructing them."[80] Regarding "the great divisions of nature," Denton concludes that neither alternative justification has been achieved.[81]

All of the above criticisms of evolution by Johnson do not necessarily mean that some form of evolution has not occurred. Johnson does not appear opposed to theistic evolution, a prospect *a priori* rejected by traditional Darwinism. Nor can he reject evolution itself, since, from a purely scientific standpoint, one of Darwinism's unscientific characteristics is its unfalsifiability. Pointing to evolution's illogical foundations and the existence of empirical data that appear to oppose the theory will likely prove insufficient to effect its demise. Darwinism turns out to be curiously like the ill-reputed scientific creationism so often decried by evolutionists themselves.

I will suspend judgment about the definitive validity of evolutionary theory for purposes of the present investigation. Speculative prudence dictates this approach because of (1) the inherent limitations of the philosophical method respecting a natural science theory and (2) the analogous limitations found in the method of experimental science.

I will now proceed by operating on the supposition, solely for the sake of argument, that some sort of biological evolution has taken place. My task in the following chapters will be to examine various philosophical and theological problems that arise in evolutionary theory, testing in each case the intelligibility and inferences which would follow.

The ongoing great debate over the general scientific validity of biological evolution and the exact form of its occurrence will be left to others. My concerns and perspectives in the present work are primarily philosophical.

Two

THE CONCEPT OF NATURAL SPECIES

Traditional Christian philosophy maintains that we can know God's existence by unaided natural reason.[1] God creates and conserves in existence all finite beings. God also sustains creatures in all their naturally proper and perfective activities. The principle of sufficient reason demands extrinsic causality by preternatural or supernatural agents to explain any activities that exceed creatures' natural powers.

By generating new and higher forms of life, evolution apparently violates the principle of sufficient reason. That principle says that a lower power cannot generate a higher. Materialists, such as Ernst Haeckel, would deny that evolution entails any such ontological ascent, and affirm that it requires no higher cause.

Those philosophers of nature who recognize in evolution progressive diversity in matter's organization and a real tendency toward higher states of being see need for a higher cause. As a Christian philosopher, I must investigate whether (1) inter-specific transformism occurs and (2) emergence of new forms entails ontologically higher states.

Anti-evolutionists generally accept "variation within the species" or intra-specific evolution, for example, Austin M. Woodbury. Biological and metaphysical objections to evolution arise only when we allege a form has transcended its species. For this reason, I need to determine the precise meaning and extension of the term "natural species." My concern is primarily limited to (1) whether natural species exist and (2) how inter-specific transformism, if it occurs, can be compatible with traditional philosophical principles.

Nominalistic denial of any extra-mental basis for species is an extreme position on the biological-species question. Common sense belief that every taxonomic sub-species, such as the domestic cat, constitutes the basis for an extra-mental, immutable essence is another extreme. This belief is almost everyone's pre-reflective view. Evolutionary biologists often embrace nominalism because they think species terms are only convenient words to describe mid-ranges of ever-blending series of unique individuals.

Extreme nominalism solves the species evolution problem by eliminating any material species to transform. It ignores what happens when we affirm organic species' extra-mental existence. My problem does not concern extra-mental species as living organisms existing in a Platonic-like world of independent forms or essences, or angelic beings. St. Thomas Aquinas maintains that angels lack matter: the principle of limitation to allow multiplication of individuals belonging to the same species. So, each individual angel constitutes an extra-mental species. My problem involves organic species, not Platonic forms or angelic species. Organisms exist as individual material beings. My concern is how we obtain a proper concept of organic species from extra-mental reality.

John N. Deely observes that the concept of species is a mental being, a being of second intention.[2] Obviously, a mental being is not extra-mental. My problem is (1) whether we abstract the intra-mental concept of organic species from some extra-mental foundation and (2) what might be that foundation's precise nature.

The biological concept is my starting point for the extra-mental concept of species. Biologist Ernst Mayr categorically rejects the traditional notion of species based upon evident morphology or physical type. He argues that the notion of "unchanging essences" makes thinking in evolutionary terms virtually impossible. Mayr insists that the evolutionist cannot think in terms of Plato and Aristotle's "typological philosophies."[3]

Mayr moves evolutionary biology away from (1) traditional classifications of living organisms based upon anatomical structure and (2) the classical Greek philosophical notion of essential differences between species. He views morphological differences to be of "strictly secondary" significance.[4] The biological reasons for his conclusions are irrelevant to my study.[5] His alternative proposal is relevant: the modern and widely accepted biological concept of species. Mayr and most contemporary biologists use the term "biological-species concept" to refer to a population system that can interbreed and have "reproductive isolation" against others.[6]

Modern biologists tend to replace the term "species" with the term "population." They no longer use anatomical likeness as their realist criterion for classing organisms together. Instead, they conceive a population in terms of group members' ability to interbreed, no matter what reason members cannot breed with other "populations."

We cannot apply the biological-species concept to asexually reproductive organisms. By definition, reproductive isolation is a property of every asexual organism. Mayr cannot explain the meaning of the term "population" among such organisms.[7] Still, he defends the validity of the modern biological-species concept, suggesting that asexuality "is almost certainly a secondary phenomenon."[8]

The modern biological-species notion also poses a problem for evolution's paleontological evidence. Fossil records do not directly preserve reproductive isolation and interbreeding, the criteria for establishing a population. Nevertheless, Mayr defends such evidence's worth by making circular appeal to "morphological criteria," which he links to "ecological, stratigraphic, and distributional evidence"[9]

Biological arguments, as such, lie outside my domain. The epistemological status of the biological-species concept does nót. Despite its various empirical advantages, the biological-species concept offers no assurance that it constitutes a valid basis for true philosophical species. The biologist seeks perinoetic knowledge: knowledge in which, by purely descriptive substitute signs, we conceptually grasp an organism's sensible or common accidents. The philosopher pursues dianoetic knowledge in which, by careful examination of essential properties or proper accidents, we understand a thing's essence or nature.[10] We know things' natures through their activity. To distinguish philosophically diverse species, we must use activities that manifest true essential distinctions. The biological-species concept offers no such evidence.

Genetically independent organisms might well possess the same substantial nature. Nothing prevents two such species from exhibiting the same proper accidents or properties. Species exhibiting many similar sensible accidents might even appear able in principle, but not in fact, to interbreed. Mayr offers an example of a species pair which exhibit such "practical problems." He comments that, in some cases of closely related species pairs, the potential mates are within "cruising range" of each other during the time of mating and, apparently, could "freely interbreed," but for some "specific isolating mechanisms."[11]

The concepts of reproductive isolation and interbreeding represent activities based solely upon the proper accident of reproduction, an essential attribute (operative potency) common to all living organisms. Such a common attribute manifestly cannot essentially distinguish any one living species from another.

The differentiating aspects of the biological species' definition pertain solely to the relative object of the mating activity (that is, with members of the same population and not with members of other populations). A common accident of relation grounds such distinctions. This relationship to another stands as a substitute sign for what biologists hope to be an ontological essential difference. It might not be. The biological knowledge is substitutional, not essential, in nature. It fails to reach the level of natural species in the proper philosophic sense.

When Mayr rejects the species concept based upon typology or morphology, he fails to distinguish between proper and common accidents taken as signs of intrinsic nature.[12] As a biologist, he tends to operate at the level of perinoetic intellection. Mayr thereby fatally flaws his attack upon the *eidos* (idea) of Plato and Aristotle. The methodology proper to the philosophy of nature alone gives genuine dianoetic knowledge of natural species.

Any typology based upon such sensible accidents as vocalization, configuration of spines, shape of mouth, number and placement of eyes, presence or absence of a backbone, and so forth, is descriptive by definition and cannot reveal the essential nature of anything.

As scandalous as this might sound to the professional biologist, search for essential natural species must transcend the methodology proper to the biological sciences. The limitations inherent in perinoetic intellection, the proper intentional act of biological science, mean that the quest for essential distinctions between living organisms is actually meta-biological, beyond biology. This quest requires the methodology of the philosophy of nature.

This conclusion might appear unreasonably deferential to natural philosophy. But Mayr affirms the need to penetrate past empirical terms, such as "phenotypic, morphological, genetic, phylogenetic, or biological" to get to the "underlying philosophical concepts," to achieve a proper analysis of the "species problem."[13] The eminent biologist invites us to penetrate beyond the perinoetic to intellection's dianoetic level if we wish to find an effective solution to the problem of biological species.

Scientist and philosopher Raymond J. Nogar moves us somewhat away from the viewpoint of the pure empirical scientist. He admits the difficulty of finding a definition of natural species satisfactory to everyone, scientists and philosophers.[14] Nogar insists upon natural species' extra-mental reality. He despairs of logic's or metaphysic's utility in defining them. His own approach is by compilation of "fundamental characteristics." Instead of trying to obtain "an essential, metaphysical definition," we must produce a "group of associated fundamental characteristics."[15] A "complex of characteristics," taken as a sign of some real unity co-ordinating the complex into a sensible whole, designated as, say, a "cat" or "dog," expresses a species' definition.

Conjoining a number of such basic attributes allows us to compare and distinguish a given type of organism from all other living types so that the *nature of natural species is sufficiently known to us through their characteristic morphology, physiology and ecology*."[16] Nogar offers a chart listing the multiple characteristics of the domestic cat so that we can clearly distinguish it from all other species. The characteristics listed include the basic powers of self-development and sensation, and several less basic attributes, such as having a digestive tract, spinal cord, skull, jaws, claws, teeth, and so on.[17]

Nogar shows no awareness that we may divide his list into proper accidents (self-development and sensation) and common or sensible accidents (the rest of the list). All animals share these proper accidents. Considered by themselves, such proper accidents would not define the house cat as opposed to other animals. When we take the remaining complex of sensible accidents as a whole, they appear applicable to the species alone. Yet, their probative force only exists when we accept that the domestic cat is actually a distinct natural species!

Such reasoning commits the fallacy of *petitio principii*. These sensible accidents converge to define no "stable, truly long lasting species." Supposing evolution, the cat blends through innumerable transitional links with other organisms.[18] Such accidental characteristics would be merely a tempo-rary cluster of distinguishing qualities whose uniqueness would be as intrin-sically transitory as the fleeting individual itself. The perinoetic character of our knowledge of such sensible accidents betrays any assumption of ontolog-ical significance.

Nogar's entire treatment of natural species waffles on their exact metaphysical status.[19] He sees species as necessary to his general defense of evolutionary theory. He supports Mayr's insistence that, without some no-tion of "type," we have no thesis about species' evolution.[20]

Nogar wants to say that the "group of associated fundamental charac-teristics" indicates a "physical essence" to which, in principle, a logical or dialectical definition with genus and specific difference could be assigned.[21] But he insists that we cannot view natural species in terms of the "*sic et non* division of cosmic reality*" that philosophers seek.[22] Nogar suggests that natural species possess ontological structures amenable to essential defini-tion. Still, he maintains that, "because of the great potentiality of the cosmic materials," we cannot achieve defined knowledge of their natures.[23]

Nogar has good reason to believe in the existence of natural species. But he admits that we cannot attain the certitude of essential distinctions for the "natural species" he has in mind. Not a remarkable result. Nogar attempts to use the biologist's perinoetic instrumentation to build a dianoetic conclusion about the essence of natural species. This weakness in Nogar's proposals reveals that metaphysical analysis from sensible experience of biological entities alone, not biological science, can produce certitude about natural species' division.

While philosopher-theologian Woodbury approaches the concept of natural species from a perspective more philosophical than the preceding thinkers, he still draws heavily from biological sources. Woodbury distin-guishes between perspectives of the natural scientist and the philosopher. The experimental scientist considers systematic species. The philosopher seeks to determine the origin of the distinction of organisms related to natu-ral species.[24]

As a philosopher, by "natural species" Woodbury means a "specific type" that only undergoes accidental differentiation.[25] By "systematic spe-cies," he understands a division that experimental scientists make for con-venience.[26]

Interpreting Woodbury's notion of natural species is difficult. He sug-gests a philosophical natural-species concept expressible in biologically equivalent terms. What experimental scientists call a distinct "class," Wood-bury understands as constituting a distinct natural species. He finds no evi-dence for the "transmutability of this type" (class) in present or past time.[27] Woodbury takes stability of organization as a "property in the strict sense" that manifests a specific essence. He maintains that, except for the vertebrates, throughout paleological time, all "supreme types of animals" have remained unchanged with respect to their fundamental organization.[28]

Empirically, we have difficulty faulting Woodbury's position because (1) pre-Cambrian fossils are scarce and (2), in the Cambrian period, most invertebrate phyla appear fully formed. Nogar observes that family relationships between various invertebrate phyla remain a puzzle even to the present day.[29] Most examples of classes remain beyond range of transitional paleontological verification.

Almost half a century has elapsed since Woodbury wrote on this subject. Today scientists dispute his claim's factual basis that classes show no evidence of evolution. Roger J. Cuffey maintains fossil records show many examples of transitional fossils. He insists that these transitions connect "low-rank taxa (like different species)" and "high-rank taxa (like different classes)."[30] Cuffey defends evidence of transitions between vertebrate classes and refers to more recent, extensive studies of reptile-mammal class transition by Beerbower (1968), Colbert (1969), Cuffey (1971), Olson (1965, 1971), and Romer (1966, 1968).[31]

Even if we grant that extrapolation from limited vertebrate data ground belief in transitions between most classes, if paleontological evidence for genuine evolution between any classes is available, the biological claim that such transitions do not occur is flawed.

To the present day controversies rage about the exact status of isolated intermediates, like *Archaeopteryx*. Doubts that recent studies pose to Woodbury's paleontological claim manifest the danger philosophy faces when it attempts to find its foundation in current and often changeable perinoetic scientific belief.

Woodbury was a philosopher and theologian. His philosophical reasoning, not his biological data, offers greatest credibility. His conclusions about class flow from metaphysical and cosmological principles and paleontological data. Woodbury bases his proof on the premise that a being's metaphysical essence roots real properties (*per se* accidents). He argues that, since their properties differ, human beings, brute animals, and plants differ according to natural species and specific essence.[32]

Intellect and will are distinctive human properties. Sensitive powers distinguish brute animals from plants. Woodbury makes essential distinctions according to properties. He maintains that things' inequality of formal perfections separate natural species.

Metaphysically, Etienne Gilson describes the progressive series of more perfect species we find in the world. He argues that an essential inequality in a thing's being generates the formal distinctness that accounts for things' specific distinction. Forms that determine the nature of various things constitute them in "different amounts of perfection." The species of natural things are "arranged hierarchically and ordered in degrees." Each level of being, from element to compound, plant, animal, and human being, is more perfect in being than the preceding one.[33]

Gilson reflects St. Thomas Aquinas' thought. Aquinas maintains that hierarchically-ordered possession of perfections distinguish natural species. He argues that human beings are more perfect than brute animals and that brute animals are more perfect than plants because "species and forms differ from one another as the more perfect and the less perfect."[34] St. Thomas cites Aristotle's *Metaphysics*, Book 7, where the Stagirite "likens the species of things to numbers, which differ in species according to the addition or sub-traction of unity."[35]

Formal distinctions, denoting ever-greater perfections, are of quality or kind. Degree of possession of rationality does not determine the distinction of human beings as a natural species. Presence of any rationality at all, as opposed to its total absence, distinguishes human beings from brutes.

Woodbury presents an argument to establish biological classes as natural species. His major premise argues that diversity of property, for example, a type of organization stable and constant under all circumstances, manifests diversity of specific essence.[36] His minor premise maintains that biological classes constitute "types of stable organization" constant through all geological history.[37] Most contemporary scientists rebut this empirically-derived minor premise.

Had Woodbury believed in fixity of natural species, as did Carolus Linnaeus, based upon Woodbury's major premise, likely he would have considered lower species to be natural species. Woodbury realized that fossil records show transitions between these Linnaean species. He granted the probability of their evolution, that is, evolution only within the same natural species.

Woodbury offers us several arguments in favor of evolution within the same natural species. For example: (1) We can directly observe that external conditions, such as food, climate, and so on, sometimes produce permanent changes in an organism's type. (2) Sometimes evolutionary process can best explain paleological evidence. Some fossils apparently show that the animal called "hipparion" emerged into the modern horse. (3) Apparently evolutionary process best explains some bio-geographical facts. And (4) experimental evidence exists to indicate that heredity can transmit factorial and chromosomal mutations.[38]

Had Woodbury been convinced that genuine transitional forms exist between systematic classes, I assume he would have denied that systematic classes are natural species. The fallibility in Woodbury's conclusion that biological classes constitute natural species lies in the contingency of his biological data, not in any weakness in his philosophical principles.

Woodbury has two arguments to demonstrate qualitative distinctions between natural species of living organisms.[39] He takes one "from diverse properties" and a second "from diverse stable type of organization."[40]

Woodbury uses the proof from diverse stable type of organization to identify natural species with biological classes. The logical fallibility of this argument arises from inherent uncertainty in biological science's typical perinoetic intellective process. Proof "from diverse properties" is dianoetically intellective and properly metaphysical. It can produce objectively-grounded certitude.

Woodbury argues that essentially diverse properties demonstrate essentially diverse natures. He claims that natural species' essentially diverse properties at least distinguish plants, brutes, and human beings.[41] This argument corresponds to Gilson's traditional claim for species division.

Woodbury shows that evolutionary ascent, in which parents lacking certain sense powers would procreate offspring possessing such sense powers, involves significant metaphysical objections. He enunciates a key metaphysical argument about natural species diversity.[42] Woodbury maintains that diversity based upon possession or lack of a sense power is diversity according to an essential property. The just-described situation means that such offspring would be of diverse natural species from their parents. This must constitute a diversity according to specific essence.[43]

An implication thus arises that we can distinguish organisms' natural species in accordance with presence or absence of different sense powers, provided that degeneration or regression does not cause a power's absence. Apparently the argument does not apply to plants or human beings. By nature, every plant must exhibit nutrition, growth, and reproduction. And human possession of intellectual faculties and will involves reciprocal implication. But brute animal sense powers multiply and vary according to their formal objects. A horse's specific essence includes the external senses of sight and hearing. These sense powers indicate that the horse's essence exceeds and is distinct from an oyster's specific essence. An oyster's external sense appears restricted to touch. Presence of at least several distinct natural animal species is evident because (1) multiple external and internal senses exist in different types of animals and (2) not all animals possess each of these senses.

While Woodbury might be wrong that biological classes constitute natural species, he correctly insists that (1) several, progressively more perfect natural species exist and (2) more than one natural species of animal exist. I am not concerned about the exact number of natural species. I will examine only the most important natural species (plants, animals, and human beings) to determine (1) whether such distinct species actually exist and (2) the metaphysical possibility and implications of inter-specific evolution.

Three

PHILOSOPHICAL POSSIBILITY
OF INTER-SPECIFIC EVOLUTION

Earlier I presented comments by thinkers such as Raymond J. Nogar, St. Thomas Aquinas, Etienne Gilson, and St. Augustine indicating that fundamentalist creationism (1) violates the principle of natural economy and (2) detracts from divine perfection by denying full causality of secondary agents. Sequential creationism suggests that God directly created the first members of natural species *ex nihilo et utens nihilo* (out of nothing and using nothing), or that He used pre-existent matter having no proximate disposition for the form He would later give it, "the slime of the earth" theory.

Theistic evolution suggests that new natural species arise from proximately-disposed pre-existent matter. Natural processes alter matter's organization to be receptive to new and higher forms. Metaphysics can explore precisely possible evolutionary elevation to higher forms. Such explanation seeks to preserve natural economy by respecting the full, secondary, natural-agent causality that allows and enables genuine transformism to occur.

Scientific creationism and theistic evolution need divine intervention. In the created order, sequential creationism requires general disregard for secondary causality's continuous generational flow. Theistic evolution respects the natural order of procreative agents within the limits of metaphysical possibility. Those limits might still arouse criticism.

Austin M. Woodbury objects to a form of theistic evolution, "monophyletic evolutionism." It maintains that "all living bodies are hereditarily connected through a common ancestral stock."[1] He adds that, at least with respect to the body, some biologists would apply this concept to human beings.

Woodbury rejects such inter-specific evolution as metaphysically impossible, except through (1) miraculous divine intervention or (2) monophyletic phylogenetic teleologism. The teleological alternative suggests a finality in living things that would lead toward higher types of organisms. Woodbury considers this type of evolution improbable.[2] Later, I will discuss this notion in greater detail.

Based upon paleontological evidence, Woodbury considers inter-specific evolution counterfactual. Most contemporary paleontologists reject Woodbury's biological data. Later, I will examine three exceptions he allows to his negative metaphysical conclusion.

In denying the possibility of monophyletic evolutionism, a philosopher should avoid the unenviable position of the eminent mathematician who, the same year that the Wright Brothers journeyed to Kitty Hawk, claimed to have demonstrated the impossibility of heavier-than-air powered flight. Still, biology's case for evolution is such that we might never grasp apodictic certainty of the historic fact.

Woodbury's understanding of intermediate forms, defined in terms of matter's intermediate organization, makes any form of true inter-specific evolution impossible to prove. Woodbury insists that no evidence of intermediate organization exists. Citing Vialleton, he insists this organizational type constitutes a specific property in a strict sense, not a formal type determined by major structure, size, and shape. Woodbury points out that some evolutionists propose *Archaeopteryx* "as an intermediary type of organization between reptiles and birds."[3] He maintains it is not intermediary. It "belongs to one clearly and completely determinate type."[4]

Woodbury argues that strictly essential organization of an intermediary type must consist in the micro-organization of the germ cell, something so small as to escape detection even under the microscope.[5]

Contemporary biologists might demur. They might indicate that the modern technique of electronmicroscopy reaches into the germ cell. Interconnectedness of all life gains startling, if not metaphysically-certain, support from amazing parallels and analogies in the DNA and RNA of all life forms, knowledge of which we have gained partly by that same microscopic technology developed subsequent to Woodbury's 1945 manuscript.

Woodbury has virtually defined out of existence intermediate forms as evidence for evolution. Contemporary paleontologists would still point to such forms as proof for inter-specific evolution, even in terms of the philosophy of nature's natural species, for example, from lower mammals to humans.[6] Such contemporary claims urge on the project of evaluating the real metaphysical possibility of true inter-specific evolution.

Woodbury's basic objection to most types of evolution is that, by proper generation, a lower species cannot produce a higher one. Such procreation violates the principle of efficient causality, since "an effect cannot be higher than its cause, and every agent produces a like unto itself."[7]

This argument at once proscribes any evolutionary theory that proposes inter-specific transformism without recourse to non-material agents. Woodbury conditions the above reasoning by pointing out three possible exceptions, since the metaphysician knows that preternatural agents exist.

The first two exceptions are (1) "bestowing of a higher generative power on a lower species" and (2) "special transitory elevation by special impulse to generate a higher species."[8] Woodbury objects to these two explanations of monophyletic inter-specific evolution because they both require miraculous divine intervention to enable a creature to produce an effect exceeding its natural limits.

Aquinas holds that, in the strict sense, miracles are "things done outside the order of created nature as a whole."[9] No created nature's effects should exceed its natural causality. Nor should the entire order of created nature produce new species that exceed in kind, not just degree, the generative power of any previously existing natural agent. Production of a new or higher species beyond the productive ability of previously-existing natural capacity is miraculous.

How could diverse natural species originate if they did not evolve and God did not miraculously create them? Woodbury appears not to deny God's original creation of diverse natural species. He denies such creation is miraculous. Woodbury holds that God's production of a new nature is not miraculous, just as God's original creation of the world *ex nihilo* is not miraculous.[10]

God creates all things *ex nihilo et utens nihilo*. He continues to create them by conserving them in existence. He sustains their natural operations. Because they belong to the order of created nature, we do not tend to call these things miracles.

Some natural species could not have existed from the earliest moments of the cosmos. Initial cosmic states were incompatible with the existence of living organisms. Special creation of new natural species long after initial creation appears miraculous because such newly-created additions of natural species are "outside the order of the whole of created nature" at that point in time.[11]

Woodbury's apparent alternative to the above exceptions appears to encounter the same objection he raises: the explanation requires miraculous divine intervention.

Aquinas maintains that, while only God performs acts of special creation, properly speaking, acts of special creation are not miracles. The notion of the miraculous entails something occurring outside the natural order. We should not call God's special creation of a new species miraculous in the strict sense of the term. Special creation is something beyond natural capacity, something that natural causes could not eventually produce.

Since only God's power can produce special creation, the contemporary mind considers miraculous and divine any form of special creation occurring subsequent to the world's original creation.[12]

Woodbury thinks that the origin of new species by means of special creation is metaphysically more economical than the technically miraculous. Since special creation involves a radical departure from the natural order and direct divine intervention, suggestion that it is not miraculous in nature tends to scandalize the contemporary mind. And special creation is as opposed to materialistic evolution as Woodbury's above-mentioned three "exceptions" to inter-specific evolution's metaphysical impossibility.

"Monophyletic phylogenetic teleogism" is the third and last exception Woodbury lists. "Monophyletic" means that "all living bodies are hereditarily connected through a common ancestral stock."[13] "Phylogenetic teleologism" means an "intrinsic tendential or teleological principle" exists in organisms leading them to produce new or higher forms of life. Woodbury compares it to the soul of the sperm in "digenetic generation" which is "on the run" toward an ultimate term: a new and complete organism. Such would be a soul, or life principle, which constitutes an inherent ordination to produce a more perfect soul. He distinguishes this tendency from accidental or environmental causes because this teleological principle is intrinsic to the organism.[14]

Post-Darwinian biologists tend to abhor the term "teleology." Classical metaphysics insists the finality principle is essential to all, including biological, beings' intelligible structure.[15] In the biological context, the soul, a living body's substantial form, is the finality principle. The soul constitutes life's principle and determines living matter's total organization.

"Phylogenetic tendency" is the soul's intrinsic finality to move toward the ultimate procreation of a new and higher natural species. If we must accept monophyletic evolution, Woodbury prefers this explanation to his other two exceptions because it does not require a miracle. Phylogenetic teleologism is "less removed from what is connatural."[16]

Still, Woodbury concludes that such phylogenetic evolution "more probably" did not occur, because "every species, properly understood as natural species, is a complete and perfect nature."[17] Every species is "ordained to its own perpetuation, not its own destruction by the production of another species in its stead."[18]

Except for the three possible exceptions just noted, Woodbury maintains inter-specific evolution is metaphysically impossible. He rejects each of the three exceptions as improbable, the first two because they entail God's miraculous preternatural intervention, the third because phylogenetic teleologism, though more connatural, offends against the completeness and stability proper to natural species.[19]

Woodbury attacks the same arguments he had proposed on behalf of inter-specific evolution.[20] He takes care not to render apodictic judgment. He postulates God's creative or formative intervention to explain an undetermined temporal succession in the emergence of diverse living species (natural species).[21] In his judgment, such divine intervention does not constitute a miracle.

Woodbury's thesis proposing special creation of natural species remains theoretically defensible because of (1) the theoretical possibility of absence of intermediate organized forms and (2) the certainty that we cannot obtain certitude in such empirical matters. Even Nogar admits, "God certainly could have extended his creative power in any way he wished."[22]

The whole disagreement between Woodbury and Nogar turns on what constitutes a miracle. Woodbury defends special creationism because he views it the only metaphysically possible and entirely connatural explanation. Nogar rejects special (sequential) creationism because he considers it non-natural, as miraculous.[23]

Woodbury's insistence on the intrinsic indiscernability of intermediary formal organization checkmates perinoetic analysis offered by biological science in favor of inter-specific evolution of philosophical natural species. Nogar's affirmation of evidential convergence in favor of descent with modification, even of philosophical natural species, demands that we try to understand evolution's possible metaphysical basis. Some natural scientists propose that empirical evidence suggests a natural determinism, not blind chance, grounds the part of evolutionary theory pertaining to life's origin. Theistic philosophers might interpret this as evidence of divine design. Philosophers should seek rational explanation of all genuinely possible scenarios.

The most fundamental of all inter-specific transitions would be from non-living to living bodies. Many biologists deny any essential distinction between non-life and life. They point to entities, such as viruses, which exhibit mixed properties and pose ambiguities of interpretation regarding their exact state. Charles De Koninck liked to say that philosophers need not consider problems posed in judging marginal cases. We can base judgment on examination of evident extremes, like the obviously non-living and living, for example, a rock and turtle. Philosophically considered, the qualitative distinction exists in the contrary, essential properties of each, in exclusive manifestation of transient (other-perfective) activity in non-living things and in immanent (self-perfective) activity in living things.

Materialist philosophers logically deny all immanent activity, since their atomistic worldview entails nothing but subatomic entities acting upon other subatomic entities. Philosophers who admit the existence of substantial beings above the atomic level would follow Aristotle in accepting a hylemorphic (matter-form) constitution of physical things. From their perspective, truly self-perfective actions (1) are typical of living things and (2) essentially distinguish living from non-living things.[24]

Nogar denies the theoretical possibility of demonstrating that non-life generated life. He insists that empirical evidence indicates evolution from non-life to life may well have occurred. His problem is to demonstrate that the seemingly possible actually happened. Regardless, Nogar describes the theory of biopoesis as "*a very interesting and fruitful hypothesis.*"[25]

Special creationism is not the only possible theory to deny generation of life from non-life. Some thinkers have claimed that the world is eternal and that life has always existed in organic form. Such a scenario would exclude the origination of life since there no first living bodies ever would have been. T. A. Goudge and J. B. S. Haldane suggest that evolution from inorganic to organic life might never have occurred. For example, non-living bodies and organisms always might have existed in an eternal world with stars and planetary systems endlessly seeding lower life forms, such as spores, into successive solar systems as they become habitable. Life would occur endlessly throughout time in the universe, without beginning for itself or the world it inhabits.[26]

Some materialists propose this explanation. Apparently, this explanation also describes Aristotle's universe, where living species are procreatively eternal. The frequent materialistic presumption that such a steady state universe obviates need for a Creator manifests a measure of metaphysical and, perhaps, scientific naïveté. Goudge says the steady state "view does not clash with any available scientific evidence."[27]

Since he wrote this in 1960, a growing acceptance of the Big Bang theory by most natural scientists has undermined Goudge's statement. Over vast periods of cosmic time, a process of sequential stellar evolution, the births and deaths of countless stars, might have existed allowing for the seeding process I describe above. This process might still exist. Yet, present scientific thought also envisions a past time during which an immensely incandescent state engulfed the whole cosmos in a fashion totally inimical to the existence of any living bodies. In 1978, astronomer Robert Jastrow wrote that, according to the then available astronomical evidence, the entire universe started some twenty billion years ago in a fiery blast so intense that "all the evidence needed for a scientific study of the cause of the great explosion was melted down and destroyed."[28]

In less than two decades, scientific consensus shifted to support a temporal beginning for our cosmos, once again necessitating some theory of life's origin. Goudge maintains that, properly speaking, life's emergence from non-life is not part of biological evolution. He argues, "the theory of biological evolution is limited to what occurred between the point in time at which the earliest living things existed and the present."[29] Before life began some kind of biochemical evolution must have existed so different from the subsequent biological evolution that we should doubt any use of the term "evolution" to explain it.[30]

While we might not be justified to use the term "evolution" to describe the biochemical processes that may have led up to the first living organisms, I will consider these processes because they may connect the various natural species I examine. I say "may connect" because, despite the weight of converging perinoetic evidence supporting evolution, philosophical certitude of the nature of such ancient events still eludes us.

Presume that life emerged from non-life. How could such a process have occurred? A long tradition denies special creationism and any form of divine design. Many atheists and evolutionists claim that primitive life arose "by the fortuitous or chance disposition of the non-living or inorganic matter."[31]

Woodbury relates what is, perhaps, the most popularized expression of atheistic evolutionism, the widespread belief that some pure chance might have generated life from non-life. Woodbury responds to this hypothesis of proper abiogenesis by pure chance: fortuitously disposed non-living matter is not in the same species as a living organism, nor in a higher one. Non-living matter is not a cause proportionate to an effect, a living organism.[32]

Woodbury applies the principle of causality, which entails that "an effect cannot be higher than its cause, and every agent produces a like unto itself."[33] He objects that (1) the effect in the case of proper abiogenesis or spontaneous generation would exceed its cause in power and (2) physical and chemical interaction alone could not avoid that entropy or perfect equilibrium that opposes the material organization every living body needs.

Living things oppose such entropy through the ongoing act of nutrition which constantly re-establishes tension by taking inorganic substances into the organism. Woodbury insists the "necessary tendency of chemico-physical interaction towards entropy" prevents this nutritive power from being a physico-chemical by-product.[34]

Virtually all contemporary scientific creationists think that the doctrine that life emerges from non-life violates the basic physical law of entropy. Woodbury insists that, unless God creates a pre-existing seminal force in matter, the physico-chemical tendency toward entropy precludes the ability of any human or angelic art to dispose matter's organization to educe non-subsistent souls of living bodies.[35]

Most current natural scientific judgment would deny that the overall necessary inclination of physico-chemical interaction toward entropy must entail impossibility of locally-decreasing entropy (such as would occur during the early biochemical steps leading to the first living bodies). John W. Patterson insists that life's emergence from non-life entails no violation of the second law of thermodynamics. He argues that "overcompensating increases in entropy elsewhere" may make up for "local decreases in entropy during self-organization."[36]

Woodbury places great importance on the entropy principle regarding life's origin because of matter's organizational effects on eduction of living substantial forms. Woodbury maintains that an efficient cause educes a new substantial form by changing matter's dispositions.[37]

Woodbury thinks that necessary conditions for attaining an ultimate, as distinct from a prior intermediate, disposition are crucial for formal eduction. This ultimate disposition (1) makes matter no longer apt for the previous form and (2) prepares it for a new form.[38] And the "ultimate disposition is never together with the form which is corrupted, but is together with that which is generated."[39]

If entropy involving all physico-chemical interaction excluded that ultimate material disposition or organization required for eduction of living form, devoid of assistance of some seminal force or preternatural influence, non-living beings could never generate life. Crucial, then, are Patterson's ob-

servations: (1) "Localized entropy reduction is an extremely common phenomenon in living and non-living systems alike."[40] And, (2) "spontaneous
complex reductions in entropy are commonplace in the natural world."[41]
Apparently, entropy is a surmountable obstacle to life's natural origin.

Abiogenesis entails that inorganic matter's chance disposition generates
living things. Evolutionary solutions might exist to metaphysical objections
to abiogenesis. The metaphysician might object, "An effect cannot be higher
than its cause, and every agent produces a like unto itself." John N. Deely responds that, as long as the totality of the interacting agents suffice to produce
matter's requisite organization, eduction of the appropriate living substantial
form will occur without violating the causality principle.

Deely maintains that the agencies that cause this organization are secondary.[42] Primary is, in some way, to obtain life's needed organization. Deely
claims that abiogenesis entails no violation of the causality principle and no
need exists for special divine "concursus (still less intervention)."[43]

Deely distinguishes between univocal causation, in which the cause
must always be proportioned to in its effect, and equivocal causation, in
which the cause "need not be proportioned to its effect except *per accidens*."[44] Transformism entails equivocal causality. Its metaphysical basis
rests in the mutual confluence of causes: "The principle is the involution and
mutual activation of the causes: *causae ad invicem sunt causae*,"[45] The
participants in this "republic of natures," the created cosmos, are reciprocal
causes.[46] As such, they may generate new and higher forms of being as the
"intersection of causal chains" produces reorganizations of matter in ways
quite novel from the perspective of individual reagents.[47]

Perhaps basic organization of non-living matter cannot be so altered to
generate higher forms. Any substance's substantial form is the intrinsic
constituent co-principle of a being that actualizes its matter's potencies. It
determines that matter to be of a certain specific nature or species. A world
of exclusively non-living things would have exclusively substantial forms of
non-living beings. In such a world, at every moment, all matter's organization would be unfit to the form of life. Apart from some exterior, preternatural agent, we could never find life. Proper abiogenesis appears metaphysically impossible.

Woodbury further objects that only the ultimate disposition of matter
occasions the eduction of a new substantial form. He adds, "The ultimate
disposition is never together with the form which is corrupted, but is together
with that which is generated."[48] A world of exclusively non-living things
precludes all forms ordered to the production of such an ultimate disposition,
since the activities and end of an agent cannot exceed its nature except by a
miracle. Since nothing can come to be except through some agency ordered
toward its production, following the principle of finality, no such ultimate
disposition can come to be. And no new living form can come to be. Ulti-

mate disposition and its proper substantial form must be simultaneous. Proper abiogenesis would seem to be impossible.

The solution to these objections rests in a proper understanding of how interaction of multiple causes can produce an effect contained in none of the interacting causes taken alone. Woodbury notes, "An efficient cause does not educe a new substantial form save by changing the DISPOSITIONS of the matter."[49] The total efficient causality of all the various influences on its generation produces the ultimate disposition. Eduction of form only requires proper material organization. Given a proper mechanism for its novel and appropriate organization, original creation of matter as potential to all non-subsistent forms sets the stage for eduction of higher forms. Chance interaction of multiple causal factors could well be such a mechanism.

Woodbury's first objection offers a self-evident truism that, as long as the forms of all things are non-living, all matter's organization must be unfit for life. In reply, we should note that causal interaction progressively alters matter's disposition through a continuum of intermediate dispositions leading toward the ultimate disposition. During this period of material preparation for the generation of its ultimate disposition, non-living substantial forms continue to inform all matter. When the ultimate disposition actually comes to be, it simultaneously educes the new living form. As long as natural agents can interact to produce new material organization, possibility exists for eduction of higher forms without preternatural intervention.

The second objection correctly says that, in a world of exclusively non-living things, no agent is *per se* naturally ordered toward matter's ultimate disposition: to the form of a living thing. Effects accidentally produced through multiple natural-agent interaction may fall outside those agents' natural ordination toward their proper ends. Chance interaction of natural agents might generate effects that none of them preordain. Chance events occur outside nature's intention. They happen because another agent hinders or affects their normal activity toward their end. Aquinas tells us that, while, in most instances, causes are ordered to their effects, sometimes accidental interference occurs, "whence such interference does not have a cause, inasmuch as it occurs *per accidens*."[50] The result does not follow of necessity from a "certain pre-existent cause."[51]

I do not suggest that St. Thomas abrogates the principle of sufficient reason or the law of causality. Effects have proportionate causes. Chance encounters of natural agents can generate effects outside their natural inclination. These may allow production of novel material organization suited to eduction of higher forms, including forms of living bodies. In this sense, chance origin of living things from non-living is rationally intelligible.

Chance, as a limited explanation of the origin of new living species, is critically distinct from chance, taken as a general metaphysical explanation of the whole universe's existence and order. The classical metaphysician can view with amusement the deadly combat between scientific materialists and scientific creationists over the abiogenesis question, as if God's existence depended upon the victors. The notion of chance presupposes an orderly universe, a republic of natures obeying the universal principle of finality, a universe whose intelligibility presupposes a supreme intelligent orderer.

Jacques Maritain exposes those metaphysical presuppositions when he says that chance entails the "encounter of causal series," each of whose causes are determined to particular ends. Chance "necessarily implies pre-ordination."[52] Cosmic structure capable of generating life presupposes order just as a roulette game presupposes the shape and structure of the wheel and ball.[53]

Classical philosophers maintain that we can know God's existence through all the things God has made.[54] These makings include (1) the world's created existence in which evolution perhaps occurs and (2) the evolutionary process.[55] Whether life's origin occurred by limited chance or direct divine intervention, the cosmos remains a creature dependent upon God for its initial creation, continued creation (conservation), perpetuation of its natural laws, and its overall direction to its ultimate end as determined by His omniscient providence.

Sidney W. Fox argues that experimental evidence today supports a position that life arose through non-random processes. He concedes that many evolutionists have assumed an "implicit belief in chance processes and random matrix." He insists that "is not a component tenet in the most modern theory derived from experiments."[56]

Fox refers to a recent re-interpretation of life's origins in terms of a natural stepwise directionality in living bodies' biochemical foundations. He defends a process which he terms "molecular determinism." Primordial reactants, not chance, form amino acids. These polymerize to form "protenoids" or "thermal proteins." Thermal proteins contact with water to form micro-structures within which is formed the genetic apparatus essential to reproduction of all living bodies.[57]

Fox claims experimental evidence supports completely natural eduction of cells from proteins, thereby completing transition to actually living organisms.[58] He insists that experimental evidence supports the surprising conclusion that proteins existed prior to DNA formation.[59]

More startling is the logical inference derived from non-random origins to the effect that life must have begun more than once! Fox writes, "The experiments suggest, moreover, numerous repetitions of the same generative process on the primitive Earth, all alike due to molecular determinism."[60] Fox even suggests that the similitude of the initial living organisms arose from molecular determinism and replication of like biochemical processes, not from actual inheritance. He maintains that the "internal self-limitations of molecular stereo-specificity" caused the same types of processes to recur "innumerable times," producing similar products. In short, life arose in like fashion multiple historical times.[61] This line of reasoning leads many astronomers and other natural scientists to believe that, whenever appropriate environment and nutrient elements occur, we will find organic life throughout the entire cosmos.

Fox thinks his theories entail no special creation or divine design. Life arises as the most economical outcome of natural molecular interaction.[62] He replaces the "mythology of Genesis" with the natural "directive nonrandomness" of chemical interaction.[63] He accepts an imperative of biochemical structure as the mechanism of self-explanatory abiogenesis, obviating the need for any such hypothesis as a Divine Designer.

Whether life arose in the cosmos by chance or molecular determinism, from a metaphysical perspective, need for a transcendent First Cause remains. Fox's terms "self-explanatory" or "self-organization" cannot legitimately indicate escape from dependence upon extrinsic causes of the complete order of being, or from the all-encompassing divine providence. The metaphysician can recognize this fact even if the positivistic scientist cannot.

Fox declares that the process of stereochemical selection is non-random and not by chance. Evolution through pure chance was never a legitimate possibility in any case, despite its evident popularity in some quarters. As a purportedly complete explanation of anything, including evolution or the world itself, chance is absurd.

Jacques Maritain considers the popular notion that total chance could somehow explain the world, "however slight the probability." He states, "An effect can be due to chance only if some datum aside from chance is presupposed at the origin."[64]

Chance always presupposes a pre-existent order and structure of those elements that construct it. "Random" dice toss presupposes the existence and proper structure of "regulation" dice and other pre-existent conditions necessary to that particular chance event. Fox's scenario presupposes the given atomic structure of physical reality and its specific potentiality to a life outcome upon interaction of its elements.

The process of non-random life development presupposes a context of random interaction of the natural agents that originate this process. Fox's thesis suggests that, while we need chance interactions to set the stage for the ordered step-wise process to start, whenever appropriate conditions happen to occur in the cosmos, statistical probability assures this progression will initiate repeatedly. This progression's non-random character appears no more remarkable than that of any other chemical chain reaction that fortuitous conditions can support, except it results in living organisms.

Describing this process as non-random is crucial to probability estimates of successful abiogenesis. Random models beget estimates of vast improbabilities. Non-random ones prompt claims of virtual inevitability.

The metaphysical principle that nothing moves itself primarily demands that we understand "self-organization" in a limited sense.[65] Fox argues, "Since the order in the amino acids results from no materials other than the reactants, *mixed amino acids must order themselves*."[66] Polymerization of amino acids under conditions like those in nature that produce protobiotic proteins are no more self-ordering than is production of sodium chloride from the interaction of sodium hydroxide and hydrochloric acid. Determinism of the process is the natural result of the interaction of the involved reagents, not of self-organization. The reagents act deterministically upon each other. Because many agents compose the solution of amino acids, the process does not violate the principle that nothing can reduce itself from potency to act.

Because an amino acid solution is many things, not one, the multiple reagents of the solution are not self-ordering. These multiple molecules order each other according to their nature's chemical finality. Given the internal necessity of their chemical structures, they interact predictably.

Stereochemical selection is a chemical expression of natural selection. Blind chance produces elemental structures that progressively perfect themselves by weeding out combinations unfit for survival. R. F. Baum maintains that, while natural selection may account for the preservation of favored species, it does not explain their origination.[67] So too, stereochemical selection may explain why chemical combinations ordered toward life through self-organization perpetuate themselves in existence, not the reason matter's nature possesses the potentiality to attain progressively higher states of being. Potency is an order toward an end. If matter did not have potency to life within its very nature, life could never arise.

Fox insists "that natural phenomena be repeatable."[68] He claims that "experiments suggest that protoreproductive cellular systems began innumerable times in uncounted locales."[69] Fox implicitly affirms that an ordination toward life belongs to matter's nature. The material universe's nature is such that, given the proper conditions, life will automatically arise.

Complete chance cannot cause chemical entities to self-organize, in nonrandom fashion innumerable times. Such repeatable phenomena bespeak characteristic activity whose sufficient reason must be in this material universe's nature. If Fox is correct, a property of our cosmos is to produce living organisms. This cosmic process uniformly and regularly generates increasingly higher forms of living bodies. "Teleological," an unspeakable term to most modern biologists, is written boldly upon it.

In the contemporary context, any attempt to advance the notion of teleologism tends to raise suspicion. Still, an innate cosmic tendency to produce ascendancy of living organisms, whatever may be its ultimate explanation, is a biological teleologism or "bio-teleologism." Such a cosmic phenomenon requires explanation for (1) efficient causality required to produce higher forms of being than previously existed and (2) finality directive of their production. No one or group of finite agents may fully contain this finality. It might even find its expression in the total confluence of that republic of natures whose ultimate end only God knows.

At this point, I do not wish to endorse the evolutionary claims put forth by Fox and others. Yet the implications of his thesis appear clear. Existence of an authentic schema of biochemical steps leading toward life argues strongly in favor of its eventual completion. Except in terms of life's ultimate fruition, God's purpose in creating an incomplete framework might appear unintelligible. Would this then mean that such abiogenesis must be understood in terms of the monophyletic phylogenetic teleologism Woodbury considers most acceptable of the possible forms of evolution, even in terms of the preternatural (non-connatural) character he finds in it? No.

Woodbury's objection to phylogenetic teleologism rests upon his claim that every genuine natural species "is ordained to its own perpetuation, not to its own destruction."[70] He argues that true phylogenetic evolution probably has not occurred. Such would require "the character of a tendential being or being 'on the run' towards something beyond itself."[71]

Chemical combinations of elements "tend towards...cessation of their own species" in forming a compound.[72] Still, they do not violate the principle that a complete and perfect nature cannot tend toward its own destruction. Woodbury argues this because chemical components seek their own destruction accidentally, not essentially. Each component "in acting on the other is consequently acted upon by that other."[73]

The molecular determinism of Fox's abiogenesis is the same. Each biochemical step toward life entails a chemical reaction in which the composing factors mutually interact. They seek their destruction accidentally, not essentially. They seek that entropy or perfect equilibrium that only completion of the chemical reaction attains. Each reagent seeks entropic discharge of its chemical activity, not its own destruction. In the process, new and diverse products form. The teleologism of non-random abiogenesis requires no preternatural intervention or force. God's original creation would instill such natural teleologism toward life in the material world.

We should distinguish Fox's natural bio-teleologism from St. Augustine's "seminal reasons" (*rationes seminales*). The Bishop of Hippo conceived his seminal reasons as "a special force superadded to common matter."[74] Natural teleologism of matter to organize itself is the intrinsic tendency of chemical elements to interact according to nature. The living product of that interaction is foreseen and foreordained by God's extra-temporal providence. The components' finality transcends them exactly as in the case when hydrogen combines with oxygen to form water. From all eternity God knows and ordains the coming-to-be of water. The teleology of water's components entails no special preternatural intervention on God's part. That teleology is contained in the intentionality of the creative act by which God gives all secondary matter its material potentialities and formal dispositions. God foresees the natural effects of creatures and creates a world where created natures naturally give rise to life.

We should not confuse such natural teleologism toward life with the philogenetic teleology of thinkers like Cardinal Mercier, Pirotta, Kolliker, and others who conceive of intrinsic tendencies and forces that God implants that operate "independently of any influence from extrinsic causes, such as environmental factors."[75] Molecular determinism's abiogenesis presupposes the critical influence of environmental factors and depends upon them for the creation of biogenetic reactions.

Natural bio-teleogism would be an instance of extrinsic finality. This means that God pre-ordains finite agents toward accidentally-ordained ends that they attain as they interact according to their own intrinsic finality. What appears as chance to some and molecular determinism to others would be, from God's perspective, the eternally foreknown process of generating a world of living things. This hierarchy of life is pre-ordained by God in the initial creative act forming the republic of natures called the cosmos.

Over fifty years ago, Brother Benignus Gerrity read bio-teleologism as consonant with modern chemically determined abiogenesis. He suggested the necessity of life's emergence. He argued that, according to natural science, matter became alive and life evolved to its present many forms because "it had to."[76] Gerrity maintained that Aquinas and natural science say the same thing from different perspectives. He cited St. Thomas to show how Aquinas views primary matter as an "appetite or urge to live and ultimately to live on the highest possible level, that is to say as the body of man."[77] The cited text reads, in part:

> Whence it must be that the appetite of matter by which it seeks a form tends toward the last and most perfect act which matter is able to attain, as to the last end of generation. However, in the acts of forms, certain grades are to be found. For primary matter is in potency first to the elemental form. While existing under the elemental form, it is in potency to the form of a compound, on account of which elements are the matter of a compound. However, considered under the form of a compound primary matter is in potency to a vegetative soul–for the act of such a body is a soul. Likewise, the vegetative soul is in potency to the sensitive and the sensitive to the intellective.[78]

And again:

Therefore, the last end of all the grades of generation is the human soul, and in this matter tends to its ultimate form. Therefore, elements are for the sake of bodily compounds, and these are for the sake of living things, in which plants are for the sake of animals, and animals for the sake of man. For man is the end of all generation.[79]

Gerrity admits that St. Thomas did not try to provide a "metaphysical ground for the evolutionary process."[80] Still, the texts cited make evolution "intelligible."[81] They show that "matter is urge to live: life is the end of matter."[82]

To offer St. Thomas's text as an explanation of modern evolutionary theory would be erroneous. Thomas apparently intended to manifest matter's appetite for more perfect forms. Aquinas indicates how prime matter's actualization by each successively higher form proximately disposes it to the next, higher form. A stepwise transition from non-life to life was remotely the object of scientific speculation at the time of Gerrity's writing (1947). He supports a bio-teleologism or physico-chemical finality, through divine pre-ordination, compatible with natural extrinsic bio-teleologism. Gerrity's conception appears fully compatible with contemporary evolutionary biochemistry's molecular determinism.

Gerrity concludes that "the physico-chemical processes which science studies" are the means of nature designed to "culminate not only in life, the end of nature, but also in the inorganic conditions of life, the means to that end."[83] He maintains that "mechanical and final causality are reciprocal in nature" and that such is "the consistent Aristotelian-Thomistic position."[84]

Natural extrinsic bio-teleologism, many contemporary natural scientists, and Aristotelian-Thomistic philosophy agree that some form of abiogenesis is a legitimate possibility or even probability. That so many scientific thinkers, such as Patterson, Fox, Gould, and Sagan, give an atheistic philosophical interpretation to the same data reveals their failure to understand adequately the metaphysical context of that data.

Four

DISTINCTIONS OF NATURAL SPECIES

Essential distinctions between lesser natural species may exist. Earlier I showed that several brute animal natural species exist. We must predicate the term "animal" generically, not specifically. Now I need to show that (1) an ontological division exists between plant and animal kingdoms and (2) this division is of qualitative diversity and degree.

Austin M. Woodbury argues that division between plants and animals rests upon a *sic et non* (yes and no) regarding sensation.[1] We say that animals possess sensation while plants do not. Some thinkers have asserted that all living bodies have sensation: Empedocles, Plato, and the Manichaeans (in ancient times) and Schopenhauer, Hartmann, and Paulsen (in modern times).[2]

In three ways Woodbury shows that plants possess no useless powers: (1) Natural selection affirms non-functional adaptation would be self-contradictory. If a mutation serves no useful purpose (lacks survival value), it does not survive. (2) From induction biologists recognize that all organs of organisms now serve, or have served, some useful purpose. Even where no apparent usefulness exists, uselessness is never evident. (3) The metaphysical finality principle necessarily implies that nature produces no powers without purpose. A natural agent cannot act without seeking its perfection.[3]

Not even plants possess useless powers. The power of sensation would be useless to plants. For at least three reasons plants can achieve ends of vegetative powers of nutrition, growth, and reproduction without sensation or any spontaneous movement: (1) Nutrition involves chemical selection, osmosis, metabolism, photosynthesis, storage, and so forth. None of these requires sensation. (2) Consequent upon nutrition and without sensation, growth occurs automatically. And (3) reproduction entails elaboration of germ cells, fecundation, and even sexual congress (by bees or wind and so forth) without sensation. Thus, vegetative life activities do not involve sensation.[2]

Plant movements, such as heliotropism, geotropism, and movements of spermata, appear to entail response to environmental factors. Biologists know these movements do not involve sensation, but are merely analogous to reflex responses that occur in brute animals.[5]

Woodbury distinguishes plant irritability from animal authentic spontaneous movement. Irritability constitutes vital activity manifesting internal finality, automatic in character and invariably predictable. In spontaneous movement, animal sensitive appetite varies reactions continuously to complex and changing sensory stimuli of the infinitely complex and variegated world.[6]

Rémy Collin maintains the difference between irritability and sensibility is analogous to automatic and spontaneous movement. The honey bee's flight manifests spontaneous movement when it "at every instant modifies its itinerary...because it experiences successive sensations which it obeys by correlatively varying its motive activity." The bee's flight path is unpredictably spontaneous because it belongs to a bee's activity as "bound up with sensibility, complex and varied like it."[7]

Even the largest plants manifest no sensitive operations or sense organs. Because these powers and their organs are the highest and most useful in animals, they operate continuously. This invisibility is inexplicable because such operations and organs can be detected in even the smallest animals.[8]

Woodbury maintains sensation is more than electrochemical alteration within neuron groups. Subjective psychic acts termed "sensory awareness" or "sense consciousness" constitute its essence. Because psychic acts are subjective, some argue that we cannot prove non-existence of plant sensory powers and activities. This is like the taunt of the skeptic who dares anyone to prove the non-existence of invisible leprechauns sitting on a chalk tray!

Operatio sequitur esse (operation follows existence) is a universal metaphysical law. Anyone who claims a plant could possess subjective sensory experience while lacking any (1) detectable response to the appropriate formal objects of the various senses, (2) spontaneous movements, (3) sense organs, and (4) evident sensitive operations, ignores this law.[9] A being must manifest itself in activities proper to its nature. Completely vegetative nature reveals its total sensory deficit through total absence of qualities and activities that attend sensory experience in animals.

Some biologists object that some existing species exhibit such marginally different activities and properties between plant and animal life that they defy classification. For example, the inferior class of *protistae*. Inability to determine presence of spontaneous movement or sense organs reveals present inability to make a definitive determination, not absence of a demarcation line within organisms. A thing cannot at once be plant and animal, just as it cannot simultaneously lack and possess sensation. Defective observation fails to prove ontological ambiguity. Living bodies possess sensation or they do not.[10] The highest plants may imitate the sensory activities of the lowest animals (in the limited analogous fashions described above). The qualitative gap separating sentient from non-sentient organisms remains infinite.

Plants and animals belong to distinct natural species. Genuine transformism from lower plant to the higher animal kingdom is metaphysically and cosmologically possible. Recall Woodbury's objection that "an effect cannot be higher than its cause, and every agent produces a like unto itself."[11] John N. Deely responds that as long as the totality of the interacting agents suffice to accomplish the requisite material organization, the appropriate living substantial form will be educed without violating causality. Chance interactions of natural agents can produce effects that are outside such natures' natural tendencies.

According to evolutionary theory, parents might not be, in their genetic specificity, total agents of generation. When no mutation occurs, parental genetic make-up determines the same natural and biological species in offspring. Even minor mutations may produce the same result. Somewhat greater mutations may alter accidental qualities sufficiently to produce gradual (or even sudden) transformism of biological species. But the natural species may remain the same. Transformism from one natural species to another would require fundamental mutations.

In every case of novelty, whether of a new biological species or a new philosophical natural species, some accidental interference occurs with the normal order of genetic reproduction. While like begets like, the fruit of natural generation may be unlike its forebears if some alien factor or agent interferes with the normal causal processes of the genetic system. New material organization may result that educes a novel living form, since efficient causes only educe new substantial forms by changing matter's dispositions.[12]

The evolutionary genetic-mutation concept presumes that some other causal factor, beyond causality appropriate to a specific parental nature alone, enters the generation process. In alteration of DNA macromolecule genetic micro-structure, germinal material organization suffers confluence of two-fold agency: (1) parents donate original DNA form of parent species and (2) novel intrinsic or extrinsic physical causes introduce new structure into the parental DNA. The ultimate effect does not exceed the causality of the total interacting causes, since the total causation includes elements that augment the univocal parental causality. This induces a form of equivocal causation into offspring production.

Any substantial form educed from matter's potency alone will be non-subsistent, incapable of existence or activity independent of matter. Metaphysical possibility allows that purely vegetative life could generate sentient living bodies by such just-described means. We need some other explanation for the origin of substantial forms that do not subsist in matter.

At this point, two questions merit consideration: (1) Do human beings constitute a genuine natural species in the proper philosophical sense? (2) How can we explain the origin of human beings in light of contemporary evolutionary theory and sound philosophical principles? Unlike lower species, human beings have a unique, spiritual dimension: freedom. Proper analysis of these questions entails theological, biological, and philosophical dimensions.

To the evolutionary materialist, human beings are highly developed animals, indistinguishable, except for greater complexity, from lower living things: the typical reductionist mentality. Yet, virtually universal belief affirms that we possess a spiritual soul and consequent immortality. The great Western philosophical tradition offers carefully reasoned arguments supporting such belief. Leading voices in that tradition include Plato, Aristotle, Alfarabi, Avicenna, Moses Maimonides, St. Augustine, St. Albert the Great, St. Bonaventure, St. Thomas Aquinas, Duns Scotus, Descartes, Thomas de Vio (Cajetan), Dominic Banez, and recent neo-scholastics.

If the human intellectual soul is spiritual, then human beings must constitute a natural species distinct from lower animals, whose souls are not subsistent. Presence of powers not found in lesser species would establish essential distinction for the human species. Higher sense powers alone would establish a novel natural species. They would not suffice to show need for a supernatural origin.

Intellect and will are distinctively human powers. Activities peculiar to human beings alone flow from these powers. In the entire animal kingdom, only human beings exhibit arts and crafts, enjoy humor, erect great civilizations and universities, produce inventions at will, make deliberate use of tools, create new languages, and make genuine progress. Most impressive is that progress by which every civilized generation advances technologically at an exponential rate while animals repeat instinctive tasks through endless generations, ever oblivious to their lives' monotony.

Intellective powers enable us to grasp intelligible aspects that lie embedded within sensible appearances of experienced things. We abstract these aspects from matter's individuating conditions to understand them as belonging to, or predicable of, a potentially infinite multitude of like objects, not just this particular object before us.

Presence of intellect within our being opens to us the problem of species. The notion of species entails some common nature predicable of potentially infinite individuals of the same type. No brute animal displays species self-awareness. Only humans, among all animals, know that they belong to a species. The notion of species is totally intellective in character. Mere brutes do not write or read treatises on the intelligibility of species.

Animal psychologist Heini K. P. Hediger observes that animals lack the notion of species. The famed gorilla, Koko, has never studied zoology or anthropology. She "could not know she belongs to the species *Gorilla gorilla* and the human beings surrounding her to *Homo sapiens*."[13]

Uniquely human intellective activities extend to (1) the above-mentioned abstractive process, (2) judgment (including self-reflection), and (3) reasoning. By these acts human beings: (1) understand essences of things in the world, (2) know the truth that these things exist, and (3) know their own existence and nature through self-reflexive awareness of the act in which they know other things. Human beings can move from one truth to another by reasoning and attain new truths hidden implicitly in old ones. We can indirectly grasp the natures of unseen things, demonstrating to ourselves and others the reality of physical entities beyond sensory range. We can discern non-physical realms deduced from the physical by philosophical sciences, especially metaphysics. No other primate is physicist or metaphysician.

Coordinate with intellect is that human appetite termed "will." By it, we freely choose between finite goods as our nature mandates we seek the ultimate good. Formal demonstration of human free will exceeds the scope of this enquiry. But few people actually suggest that lower animals have any ultimate control or responsibility for their instincts, appetites, drives, or lives. Materialistic thinkers who deny human freedom often inconsistently triumph the virtues of a free and democratic social order, as if the whole could operate in a manner transcending the determinism of its parts!

My task at this point is to delineate between human intellective life and brute sentient life. I aim to refute sensist philosophers who reduce all human knowledge and activities to sensation and sense appetite. I need not exhaustively achieve this goal. Showing that the most sophisticated sensory activities bear no legitimate threat to the human intellect's spiritual superiority will suffice to support my case.

Five

SIGNIFICANCE OF RECENT
APE-LANGUAGE STUDIES

Lower life forms often imitate activities and perfections of higher forms. Tropisms in some plants do not constitute sensation. They deceptively simulate sensitive reactions proper only to animals. Clever animals' human-like behavior causes much contemporary confusion among many people, including presumed experts on animal behavior. Today, many people think of themselves as merely highly-developed animals.

Darwinian evolution and its attendant reductionism have succeeded in dominating natural sciences dealing with animal and human behavior. Psychologists, zoologists, biologists, anthropologists, and so forth, tend to view human behavior as an extension in degree, not in kind, of lower animal behavior. We see this tendency acutely in contemporary ape-language studies' controversies. For more than fifty years, a handful of research projects have tried to teach chimpanzees and other primates to talk. More successful techniques have involved American Sign Language and computer-based artificial language systems. Since the 1970s great publicity has attended these efforts. Claims exist that these subjects (1) understand hundreds of words, (2) invent new complex words, and (3) even form sentences with two-way conversations between trainer and primate and primate and primate.

By 1979, a simmering academic controversy about the legitimacy of primate linguistic credentials burst into public view. *Psychology Today* published two skeptical articles: one by Columbia University psychologist Herbert S. Terrace, the other by University of Indiana anthropologists Thomas and Jean Sebeok. Terrace's research project was a chimpanzee named Nim Chimpsky. Careful re-evaluation of Nim's signing activities led Terrace to conclude, "I could find no evidence of an ape's grammatical competence, either in my data or those of others."[1]

The Sebeoks argue animal researchers unwittingly deceive themselves by accepting unconsciously-cued behavior as linguistic competence. They refer to the Clever Hans effect, named after a famous early 1900s "thinking" horse. Berlin psychologist Oskar Pfungst exposed Hans' "intelligent" answers to questions as results of unintentional cues given by his questioners.

Defenders of apes' linguistic abilities engaged in immediate counterattack. They produced an intellectual battle that still rages. Almost all participants in this debate are natural scientists who agree on human beings' materialistic and evolutionary origins. Dualist philosophers and theologians have said little. Critics of linguistic apes operate largely from a perspective which (1) views humans as only highly developed animals and (2) ignores philosophical arguments for the human soul's existence and spiritual nature.

Among the apes' defenders, Suzanne Chevalier-Skolnikoff claims the famed signing chimp, Washoe, taught another chimp, Loulis, how to sign. But, she concedes, "Loulis learned his signs mainly by imitation."[2]

Chevalier-Skolnikoff claims that she has observed the gorilla, Koko, engage in deception, lying, and joking behavior, none of which can be cued. She claims that Koko has been recorded to argue with and correct others, both of which require mental representation.[3]

Intentional lying, deception, joking, arguing, and correcting, if actually demonstrable, would bespeak unequivocally intellective activity. We must be careful about drawing such inferences from available evidence. We should avoid replacing facile explanation by lower causes with more difficult explanation by higher ones.

I cannot explore and critique the data that form the basis for Chevalier-Skolnikoff's judgments in this study. I can note that such judgments necessarily flow from interpretation of the concrete details examined. Herein lies the greatest danger to the human researcher who attempts to read the animal subject. The Sebeoks insist the investigators unwittingly enter into "subtle nonverbal communication" with their animal subjects. The ape-language researchers make the mistake of assuming that the subjects' reactions are "more humanlike than direct evidence warrants."[4]

The Sebeoks describe the anthropomorphic fallacy, the error of attributing human qualities to animals based upon our nearly irresistible temptation to put ourselves in the brute's place. We then view its actions in terms of human intellective perspectives. This human tendency is so universal that experts in animal behavior frequently fail to avoid its pitfalls.

The Sebeoks describe ape researchers' habitual anthropomorphism, noting how they frequently read anomalous signing as "joking, insults, metaphors, and the like." They report how an animal that was being taught the sign for "drink" "made the sign perfectly, but at its ear rather than its mouth."[5] This suspicion strikes at the heart of Koko's claimed performance of "deception, lying, joking, etc."

Psychologist Stephen Walker thinks the synergism of anthropomorphism and Clever Hans effect justifies skepticism about all claims made for American Sign Language trained apes. He points to the inherent problem that arises every time we say apes produce meaningful sequences of signs. Such sequences nearly always occur in close interaction with a human trainer. We can always charge that human contact "determined the sequence of combinations observed."[6]

Not all ape communication research techniques involve use of American Sign Language. Some techniques employ plastic symbols, computer-controlled keyboards, and other artificial devices to lessen, or possibly eliminate, human influence. Psychologist Duane M. Rumbaugh defends computer-controlled keyboard system research of Savage-Rumbaugh. He insists evidence shows clear capacity for categorization free from any Clever Hans effect. Rumbaugh claims the symbols their apes use are "referential, representational, and communicative in value." He claims that the data show that, under conditions entirely free of human influence, their chimpanzees could "categorize learned symbols as foods and tools (nonedibles) just as they categorize the physical referents themselves."[7]

As in all other instances of supposed lower primate intentional communication, a two-fold fundamental problem remains: (1) human influence in programming animal training and responses and (2) human beings' tendency to anthropomorphize the interpretation of the influenced results. Such results always appear less definitive to skeptics than to the researchers who nearly live with the subjects they wish objectively to investigate.

The Sebeoks explain the difficulty in trying completely to eliminate the Clever Hans effect: apes require much coaxing to get them to participate in these laboratory experiments. They cite Heini K. P. Hediger, former director of the Zurich zoo, who sees the task of getting rid of the Clever Hans effect as "analogous to squaring the circle." Hediger is quoted as saying that "every experimental method is necessarily a human method and must thus, per se, constitute a human influence on the animal."[8]

The entire case against "talking" apes ought not rest upon the Clever Hans effect as championed by the Sebeoks. Stephen Walker points to research done by Roger Fouts, the Gardners (with the famous Washoe), and Savage-Rumbaugh as apparently escaping the charge of unintentional cuing. He grants the "robust" character of the evidence presented by Savage-Rumbaugh in the case of two chimpanzees using the computer-controlled keyboard system with no humans present.[9]

Clever Hans effect now includes two distinct aspects: (1) unintentional animal cuing and (2) human influence upon animals. Undoubtedly, the Sebeoks and other critics correctly insist that human influence inheres in every humanly-devised ape experiment. Still, unintentional cuing cannot explain all significant ape communicative achievements.

Given exhaustive, and sometimes exhausting, training by researchers, solid documentation exists of several novel, impressive ape communication performances, free of all unintentional cuing. These cases are more than the well-known abilities of trained chimpanzees and gorillas to (1) associate arbitrary signs with objects and (2) string together series of such signs in what Terrace and others dismiss as urgent attempts to obtain immediately sensible rewards. More impressive experimental results are now forthcoming.

Savage-Rumbaugh conducted experiments in which two chimpanzees learned to communicate and cooperate with each other using a computer keyboard to transmit information revealing location of hidden food.[10] In another experiment, after extensive training and prompting, upon primate partner request, the same animals learned to cooperate by handing over the correct tool needed to obtain food. This occurred using computer symbols and with no human presence during the actual experiment.

Walker concludes the animals produced "mental associations" between visual patterns employed and objects they represented.[11] These same prodigious chimpanzees advanced to seemingly abstract symbolic associations. Through training, they could associate labels with objects that the labels classified. They could associate those same labels with the more generic label that represented the class of objects to which the labels themselves belonged. Walker gives the example: "for instance, if shown the arbitrary pattern indicating 'banana' they responded by pressing the key meaning 'food', but if shown the symbol for 'wrench' they pressed the 'tool' key."[12]

Woodruff and Premack devised a cuing-free experiment in which chimpanzees indicated by gesture food's presence in a container to human participants. The participants did not know the container's location. They would correctly direct "friendly" humans who would then share the food with them. The chimps would mislead "unfriendly" humans who would not share the food, since the animals were then permitted to get the food for themselves.[13]

The above experimental successes might involve original human influence in the training process. Apparent freedom from the Clever Hans effect of unintentional cuing makes each above experiment significant. And each demonstrates (1) fairly complex symbol-object associative skills, (2) "intentional" communication, and (3), in the last case, some form of "deception." I place quotation marks about the terms "intentional" and "deception" because we still do not properly understand the exact cognitive content of such acts.

The above-described results of (1) non-cued experiments, (2) claims of hundreds of "words" being learned, and (3) signing apes articulating "sentences" and "dialogue" are significant. But careful natural scientific observers remain convinced that essential differences remain between ape and human capabilities.

After analysis of ape-language studies' data and arguments, Walker concludes that human linguistic capabilities remain unique. He maintains that apes trained to use artificial communications systems do not possess language in the sense that human beings do: "Human language is unique to humans."[14] He suggests that similarity between a chimpanzee and a human being using American Sign Language is "in some senses no greater than the resemblance between the speech of a parrot and that of its owner."[15]

We might train a parrot to say, "Polly wants a cracker because Polly is hungry and because Polly knows that a cracker would neutralize the hyperacidity of her stomach acid and thereby reestablish its normal pH." We might even train it to say this to obtain food when hungry. No one would contend that this bird understands concepts such as "neutralize," "hyper-acidity," and "normal pH." To associate a trained response with a given stimulus is entirely other than to grasp intellectually each employed concept's intrinsic nature and the entailed cause-effect relationships.

Aside from the sheer quantity of associations learned, apes' capabilities do not qualitatively exceed those of lower species, as when a dog responds to the arbitrary sign of a buzzer to obtain a piece of meat through performing some trained action.[16] Walker also notes animals' essential dependence upon human influence to assure their performance, as in Savage-Rumbaugh getting chimps to use the keyboard in the absence of a human being.[17]

Walker describes a radical wall of separation distinguishing human beings from all lower primates, pointing especially to our unique possession of language in its proper meaning. He notes many "discontinuities between man and animals" including "human faculties for abstraction, reason, morality, culture and technology, and the division of labor."[18] He terms language "the evergreen candidate for the fundamental discontinuity."[19]

Walker maintains that, while chimpanzees form "mental" associations, their abilities pale compared to humans. He objects that no one has shown that "one chimpanzee gesture modifies another, or that there is any approximation to syntax and grammar in the comprehension or expression of artificial gestures."[20] Thus, comparisons of apes' use of signs and people's use of words is "definitely limited."[21]

Despite Walker's defense of human uniqueness, he shares most natural scientists' tendency to describe lower primates' associative imaginative acts in philosophically misapplied terms, such as "mental," "understand," and "think." Proper philosophical use of such terms strictly limits them to human intellective activities. Their application to brute animals in this context serves to confuse intellective with sentient orders.

Hediger supports the claim by bio-philosopher Bernard Rensch, who noted in 1973 that we have discovered nothing like human language among apes in the state of nature. Hediger insists that language-trained animals are "anthropogenous animals," virtual "artifacts."[22] He says we cannot know how much of their behavior is their own and how much we have instilled "through the catalytic effect of man."[23] He terms this problem "the alpha and omega of practically all such animal experiments since Clever Hans."[24]

All ape-language studies presuppose invention of true language by true humans. We then impose this uniquely human invention upon apes. The day apes create their own linguistic system is still the dream of science fiction.

The science of philosophical psychology teaches that human language consists of a deliberately invented system of arbitrary or conventional signs.[25] The English word "red" could just as well have stood for the natural color green, except for the convention or agreement that it should represent just what it does. The alternative to arbitrary signs is natural signs which flow from a thing's nature. Thus, smoke is a natural sign of fire, a beaver slapping its tail on water is a natural sign of danger, and the various calls of birds are signs of specific natural meanings, not subject to arbitrary interchange or invention. A cat's hiss is never equivalent to its purr.

In teaching apes to talk, we impose upon them our system of arbitrary or conventional signs. The signs belong to us, not to the apes. Apes use them only because we train them to do so. We turn the apes, as Hediger says, into "artifacts" of our own creation.

Hediger emphasizes the importance of not underestimating impact of human training upon lower species. He notes that, when researchers "enter into language contact, into dialogue with apes," they hope the animals will react with "certain signs in which we would like to see a language."[26] He objects that we do not know the animals understand their answers as linguistic elements. How do we know they are not just "reactions to certain orders and expression, in other words simply performances of training?"[27]

Small children also react to training by performance without understanding. Small children speak sentences, even with perfect syntax and grammar, whose meaning utterly eludes them. At least in some cases, we hope it eludes them! If such can occur in children through training and imitation, Hediger's hesitancy to attribute intellective understanding to brute animals makes sense when we can explain such acts by performance training.

Hediger suggests a technique designed to assure that apes understand the meanings of the "words" they gesture under present methods. He says "a better clarification could be reached mainly through the introduction of the orders 'repeat' and 'hold it.'"[28] He says this would assure chimps really understand and are not just making "fast, sweeping movements" into which we read understanding falsely.[29]

Since no one has attempted such "stop action" in present ASL trained apes, demonstration of intellectual understanding of hand signs in them is virtually impossible. By contrast, human beings frequently do explicate their precise meanings to each other, even to the point of writing scholarly papers immersed in linguistic analysis.

Hediger makes a fundamental observation designed to cut the Gordian knot of the controversy about apes' supposed mental abilities. If apes are so intelligent, he asks why can't they learn to clean their own cages and prepare their own food?[30] He adds, "Apes have no notion of work. We might perhaps teach an ape a sign for work but he will never grasp the human conception of work."[31]

Hediger maintains that "the animal has no access to the future. It lives entirely in the present time."[32] And again, Hediger insists, "To my knowledge, up to now, no animal, not even an ape, has ever been able to talk about a past or a future event."[33] The Sebeoks refer to Hediger as the "world's leading authority on human-animal communication."[34]

Walker's conclusions cited above warrant special attention because his book, *Animal Thought,* is a synthesis of animal mental process data. It also reviews ape-language studies.[35]

Beyond the experimental science distinctions between apes and human beings noted above, an evidence pattern exists supportive of some philosophical conclusions. Well-documented ape activities exclusively focus upon the immediate, particular objects of sense consciousness. Apes seek concrete sensible rewards readily available in the present. Such documented observations are consistent with the purely sentient matter-dependent mode of existence specific to animals.

Apes have no proper concept of time in terms of knowing the past as past or the future as future. They offer no simply descriptive comment or pose questions about the contents of the passing world. Even small children do this when they ask their father why he shaves or tell their mother she is a good cook, even though they are not presently hungry.

Repeatedly, any ape's most pressing obsession is immediate acquisition of a banana (or its equivalent). Apes show no concern for botanical speculative inquiry about such objects. Apes' whole experiential world is so limited that researchers' motivational tools used to get them to perform or dialogue are severely restricted. Hediger laments that these tools are only things like "food and drink, social and sexual contact, rest and activity, play and comfort," and so forth.[36] Similarly, Aristotle maintains that animal life focuses on "procreation and feeding; for on these two acts all their interests and life concentrate. ...All animals pursue pleasure in keeping with their nature."[37] Small wonder apes do not philosophize or clean their cages.

We can explain many ape-communicative skills as simple imitation or unintentional cuing. Even when carefully controlled experiments seek to lessen or eliminate cuing, to get apes to initiate and continue their performance we simply cannot eliminate human influence and extensive training.

Still, impressive experiments done by Savage-Rumbaugh and others manifest sophisticated ape-communicative skills. Exhaustive training may account for these chimpanzees and gorillas acting in ways never seen in the state of nature. Training alone does not fully explain such remarkable behavior in this human-imposed artificial state.

No undisputed evidence exists of ape-language skills that exceed the association of sensible images. Categorization of things like tools and actions does not exceed the sensible abilities of lower species. Consider the ability of a bird selectively to recognize objects suitable for nest building. And the ability to label labels does not exceed, in principle, the province of the association of internal images.

Intellective knowledge consists in more than the ability to recognize common sensible characteristics or sensible phenomena associated with a given type of object or action. Whenever animals respond in consistent fashion to like stimuli, all animal species display such sentient recognition: for example, when a wolf senses any and all sheep as appetitive objects.

The intellect penetrates beyond things' sensible appearances to their essential nature. The intellect's first act (simple apprehension or abstraction) "reads within" the entity's sensible qualities, grasping intelligible aspects that it raises to the level of the universal concept. We can imagine an individual triangle's sensible qualities. We cannot imagine the universal essence of triangularity, since a three-sided plane figure can be expressed in infinitely varied shapes and sizes. The concept of triangularity is a proper object of intellective understanding.

Conceiving the universal consists in more than associating similar sensible forms: formation of a concept abstracted from the individuating, singularizing influence of matter, freed from all the sensible qualities able to exist only in an individual, concrete object or action.

A chimpanzee's correct identification of, communication about, and employment of an appropriate tool to obtain food is no assurance of true intellective understanding. A spider weaving its web to catch insects repeatedly creates the same type of tool designed exquisitely to catch the same type of victim. Such instinctive behavior hardly bespeaks true intellectual understanding of the means-end relationship on the spider's part. The moment a spider must perform any feat or task outside its fixed instinctive patterns it manifests its lack of intellect.

Nature programs the spider, human beings the chimpanzee. Each animal follows pre-programmed habits based upon recognition or association of sensibly similar conditions. No ape or any other brute animal understands means as means, the end as end, and the relationship of means to end as such. The sense is ordered to the particular; only intellect understands the universal.

How do we know that apes do not understand the intrinsic nature of objects or labels we have trained them to manipulate? Apes appear to act quite "intelligently" within the ambit of their meticulous training. Still, just as a spider cannot perform outside its programmed instincts, apes exhibit no originality or creative progress as do humans when we invent at will our languages, build great civilizations and, yes, keep our "cages" clean!

Trainers have conditioned apes to associate impressive numbers of signs with objects. Mere association of images with signs and objects, or even of images with other images, does not constitute evidence of intellective understanding of anything's intrinsic nature. Such acts of understanding remain exclusive to the human species.

Ape-language studies also center upon second and third order intellective acts, judgment and reasoning. Chevalier-Skolnikoff insists that the chimpanzee, Washoe, and the gorilla, Koko, exhibit true grammatical competence.[38] She reports Koko signing "breakfast eat some cookie eat" and "please tickle more, come Roger tickle," with Washoe also signing "you me go peekaboo" and "you me go out hurry." She claims these are "intentionally planned sequences" giving new information.[39]

She reports Koko to have argued with and corrected others using such "intentionally planned sequences." When Koko pointed to squash on a plate and her teacher signed "potato," reportedly the gorilla signed "Wrong, squash."[40]

The above-given data is anecdotal. More problematic is the inherent danger of anthropomorphic inferences that Walker, the Sebeoks, and others warn against. Walker concludes, because of interaction with a human companion during such communication, virtually no evidence of this kind exists "not vulnerable to the charge that the human contact determined the sequence of combinations observed."[41]

Precisely how trainers conditioned the animal to sign "wrong" or otherwise indicate a negative is not evident. Such a sign when associated with a correct response ("squash") need not reflect a genuinely intellective judgment. The correct response itself is simply proper categorization, the product of training. Its association with a negative word-sign like "wrong" or "no" may simply be a sign trained to be elicited whenever the interlocutor's words or signs do not fit the situation.

Presumption of intellectual reflection and negative judgments in such cases constitutes rank anthropomorphism in the absence of other specifically human characteristics. No data records a correction or argument entailing a progressive reasoning process. Instead, two signs, such as "No, gorilla," or "Wrong, squash," constitute the entire argument. Compare such simple denials to the lengthy syllogistic arguments, often of many steps, offered in human debate. Apes offer small collections of associated simple signs, usually united only by the desire to attain immediate sensible rewards.

As noted above, some researchers report that apes sign to other apes.[42] Reportedly, when alone, they sign to themselves.[43] Such behavior, though striking, reflects force of habit. Once apes well establish proper associations of images to hand signs, the tendency to respond in similar fashion in similar contexts, whether in the presence of humans, another ape, or in solitude, is not remarkable.

Terrace offers stinging defection from those advocating an ape's grammatical competence. His own chimpanzee research subject, Nim Chimpsky, led him to question initially favorable results. He started complete reevaluation of his prior data and that available from other such projects. He now insists that careful analysis of all ape-language studies fails to demonstrate that apes possess grammatical competence.

Terrace argues that two studies using artificial language devices showed what the chimpanzees "learned was to produce rote sequences of the type ABCX, where A, B, and C are nonsense symbols and X is a meaningful element,"[44] The sign "apple" might have meaning for the chimpanzee, Lana. In the sequence *"please machine give apple*," he doubts "Lana understood the meanings of *please machine* and *give*, let alone the relationships between these symbols that would apply in actual sentences."[45]

Terrace emphasizes importance of sign order in demonstrating simple constructions, such as subject-verb-object. He criticizes the Gardners for failing to publish data on sign order regarding Washoe.[46] His single most important contribution has been to obtain and analyze a large body of chimpanzee-combined signs to check for sign regularities.[47]

Terrace initiated a careful videotape examination of Nim's and his teacher's signing, so detailed that it took an "hour to transcribe a single minute of tape."[48] These careful examinations led him to conclude that apes sign in response to urgings and to obtain objects or activities, not to create particular meanings.[49] Terrace claims that to evaluate the performance of other signing apes is difficult because (1) discourse analyses have yet to be published, (2) the data is anecdotal, and (3) such data fails to give full listings of combinations.[50]

Mark S. Seidenberg and Laura A. Petitto raise similar objections against anecdotal evidence. They argue sparse data allow for multiple interpretations. Ape researchers prefer the strongest. They claim the ape signed "creatively." These vague anecdotes are the sole support for two significant claims: (1) Apes create novel sequences and (2) they create "syntactic structure."[51]

Terrace examined Washoe's and Koko's films and videotape transcripts. He concludes "discourse analysis" reveals "Washoe's linguistic achievement less remarkable than it might seem at first."[52] He also examined four transcripts' data on two other signing chimpanzees, Ally and Booee.

Terrace concludes signing's sole function for all four chimpanzees is to obtain rewards. He maintains little evidence exists showing "that an ape signs in order to exchange information with its trainer, as opposed to simply demanding some object or activity."[53]

Seidenberg and Petitto note need for a large body of context-analyzed utterances to determine linguistic competence. They conclude, "No corpus exists of the utterances of any ape for whom linguistic abilities are claimed."[54] Terrace claims no linguistic ability for Nim Chimpsky. Seidenberg and Petitto affirm data collected on the Nim project is "more robust" than other ape projects. While the data on the other four apes listed above "are limited in several respects, they are the only systematic data on any signing ape."[55]

Seidenberg and Petitto's judgment combined with Terrace's discourse analysis means that ape-language studies fall into two categories: (1) those, such as the Nim project, which are based upon systematic data, but for which discourse analysis reveals little evidence of grammatical competence and (2) the rest of the projects, for whose subjects various claims of linguistic ability have been made, but none of which are based upon systematic data.

Another weakness in the data, afflicting even the Nim project, is the practice of simply deleting immediately repeated signs. Apes tend to repeat signs. Humans do not. Researchers always eliminate this primary difference.[56] Deletion of such uselessly repeated "words" would tend to make an ape's recorded "speech" appear much more intelligible and meaningful than it is. Seidenberg and Petitto conclude that this research contains "numerous methodological problems."[57]

In more recent comments, Terrace maintains that first generation ape-language projects overlooked the importance of ability to refer with names in their zeal to emphasize the ability to generate sentences.[58] He notes that human children, unlike apes, appear to "derive intrinsic pleasure from the sheer act of naming." They refer to objects spontaneously with no interest in obtaining them.[59] Children take "sheer delight...in contemplating an object and sharing it perceptually with the parent." Apes do not.[60] While human children take natural enjoyment in the process of naming things, "whatever referential skills an ape displays naturally seem to be in service of some concrete end."[61] While apes must be arduously and endlessly drilled to learn signs, small children, once they grasp the game, spontaneously want to learn names for everything in sight. Terrace suggests that future ape-language projects focus upon possible referencing skills, perhaps, using *Pan paniscus*, the so-called pygmy chimpanzee which has shown some natural pointing behavior.

The main question is whether animals, limited to sense faculties alone, can ever understand the nature of referencing at all. Can they grasp the nature of naming? Evolutionary materialists expect sense powers to accomplish this task. Philosophers who defend the intellect's spiritual nature object that sensation alone can never understand the nature of words as such. Apes will be forever trapped in their focus for immediate, concrete sensory rewards.

What if all data from ape-language studies, anecdotal and otherwise, were accepted at face value? Claims that apes understand meaning of signs, create new word complexes, deceive, lie, reason, and so forth, still need not prove they have intellective powers. We can program contemporary electronic computers to simulate many and, in principle, all of these behaviors.

As evidence of potential to engage in "low-grade" conversations, Walker points to "computer programs which can play the part of a psychotherapist in interchanges with real patients."[62] Recent computer achievement in defeating a world-class chess champion underlines the point. Computers can surely perform many functions of signing apes.

Given appropriate sensing devices and robotics, computers could simulate even most impressive, non-cued Savage-Rumbaugh experimental results. Computer pairs could exhibit co-operative exchange of information and objects, as seen in the activities of chimpanzees, Sherman and Austin.[63] This would include the ability to label labels and respond to the arbitrary pattern for "banana" by pressing the key meaning "food."[64] While such performance might seem remarkable in an ape, it would be child's play to a properly programmed computer.

Programming a computer to "deceive" or "lie to" an interrogator is no great feat. Woodruff and Premack apparently spent considerable time and effort creating an environment which programmed chimpanzees to engage in just such unworthy conduct.[65]

No reports exist about apes having learned to play chess. Walker says pocket-sized computers can now play chess at a routine level "accompanied by a rudimentary attempt at conversation about the game."[66] Electronic computers engage in "low-grade conversations," probably outstripping their nearest ape competitors.

The above-mentioned capabilities presuppose intelligent human computer programmers. Correlative ape programming occurs through (1) deliberate human training, (2) unintentional cuing, and (3) unavoidable human influence upon animals.

Apes' capabilities equal or exceed computers in several significant respects. Physicist-theologian Stanley L. Jaki maintains the estimated number of neurons in the human brain is 10 to the 10th power.[67] He says "the human brain" has "twice as many neurons as the number of neurons in the brain of apes."[68] The number of neurons in the ape brain must be 5 times 10 to the 9th power. An impressive amount of instantaneously available "core storage and central processing capability." Elaborate sensing devices can provide input data to a computer. But no artifact matches natural abilities of multiple external and internal senses found in higher animals, including apes. Ape ability to sense and categorize bananas as food is simply part of their natural equipment. While complex robotic devices are an essential ingredient in computer-controlled manufacturing processes, an ape's limbs, hands, and feet afford comprehensive dexterity unmatched by any single machine.

None of the performances exhibited by language-trained apes exceed in principle capacities of electronic computers. Electronic computers simply manipulate data. They experience no intellective or sentient knowledge, and possess no unity of existence proper to a single substance. A computer is merely a pile of cleverly constructed electronic parts forming an accidental, functional unity to serve human purposes.

No surprise that human beings can program apes to perform as researchers reported. As Hediger points out, these apes have become artifacts through the language and tasks we humans impose on them.[69]

My claims might appear gratuitous in an age in which many computer experts proclaim imminent success producing artificial intelligence through cybernetic science. Those who fail to understand that computers possess no substantial existence and unity or any sentient or intellective knowledge will fail to grasp my above argument from analogy. Still, whether we consider them singly or in concert, elementary components of complicated contemporary computers experience nothing whatever. No inanimate substance, atom, molecule, rock, or electronic chip, can sense or understand.

What answer can we give to the skeptic's seemingly absurd, but elusively difficult, query: "How can we be certain that the apparently inanimate parts composing modern computers do not contain some form of consciousness, or at least the potency for consciousness?" Even novice logicians know well the problems of demonstrating a negative conclusion. While proving that inanimate objects are truly non-living, non-sensing, non-thinking, and so forth, is difficult, such proof is possible the moment we take the issue seriously.

Sensitive, intellective, and other life potencies only exist as faculties (operative potencies) of already living things. These powers are secondary qualities inherent in, and proper to, the various living species. Such properties (1) flow from organisms' essences and (2) are activated by apprehending appropriate formal objects. An animal's potency for sight is a sensitive faculty of its substantial form or soul. When its proper sense object, color, activates this faculty, the animal sees. Inanimate objects do not possess such potencies or faculties.

Despite apparent probative difficulty, universal absence of specific life activities in the individual and the species evidences that such life qualities transcend, or are missing from, such a nature. A given form's presence necessarily implies its formal effects. If a thing is alive, it must manifest its life activities. If sensation is a power of its nature, it must, at least at times, actually sense. That a power should exist in a given species, but never be found in act, is impossible. We can show this fundamental truth as follows.

Metaphysical certitude demands that a sufficient reason exist that something consistently exhibits some qualities or activities but not others. Assume a non-living thing, such as a rock, manifests qualities of extension and mass, yet never exhibits any life activities, for example, nutrition, growth, or reproduction. Such life powers must be absent from its nature altogether, or, if present, some sufficient reason must exist that such powers never act. That reason must be naturally intrinsic or extrinsic. If extrinsic, it would have to be accidental to the nature, and, thus, caused. As St. Thomas Aquinas observes, "Everything that is in anything *per accidens*, because it is extrinsic to its nature, must be found in it by reason of some exterior cause."[70]

What does not flow from a thing's essence cannot be universally in that thing, even if what flows from its essence is the universal absence of a quality or activity. Aquinas argues that, since every agent acting in virtue of its nature is determined to one effect, every natural agent "comes always in the same way, unless there should be an impediment."[71]

An extrinsic cause might occasionally interfere with a living thing's vital activities. Such suppression of natural activities is relatively rare, never universal. An extrinsic cause may suppress reproduction in a few individuals in a species, but most will reproduce. If reproduction were absent in every species' member, as is the case in rocks, absence of such activity must be attributed to the essence itself.

If a thing essentially possesses a power or potency to a certain act, and that essence is responsible for its never actually exercising such a power, such essence becomes self-contradictory. The same essence would be responsible for its substance's essential ability to possess that quality and for it never actually possessing that same quality. The same essence would be the reason why a thing can be alive or conscious and, also, simultaneously, the reason why that same thing is never actually alive or conscious. Absurd, impossible.

Aristotle defines nature as "a source or cause of being moved and of being at rest in that to which it belongs primarily."[72] A nature which would also be the reason for a thing not moving or resting would contradict itself.

If every member of a species lacks a quality or activity, the quality or activity is absent, not by accident or as a positive essential effect, but because it does not belong to its essence at all. Hence, non-living things have no life powers within their natures. They can gain life powers only by undergoing a substantial change. They must somehow become assimilated into the substance of a living thing, as when a tree turns nutrients from the soil into its living self.

Substantial change into a living substance does not occur when inanimate parts artificially conjoin as an accidental, functional unity, such as an electronic computer. None of a computer's inanimate, individual parts exhibit properties of life, sensation, or intellection. No combination of non-living entities, not even a highly complex functional unity, can perceive or think. Such noetic perfections transcend individual non-living natures, and computer components' natural limitations.

A computer constitutes an accidental unity, an artificial composite of many substances. No accidental perfection ungrounded in its constituent natural elements can exist in it.

Metaphysical fantasy might tempt belief that somehow the whole might be greater than the sum of its parts, the total collectivity might exhibit qualities of existence found in none of its elements. Like Pinocchio, we might aver that the computer suddenly assume properties of a living substance, to sense and think.

We base such self-creativity on the fallacy of composition. We attribute to the whole qualities found in none of its parts, like saying an infinite multitude of idiots could somehow, if only properly arranged, constitute a single genius. The principle of sufficient reason is the fundamental obstacle to such speculation. Non-living being offers no existential foundation for life properties. Accidental rearrangements of essentially non-living components provide no sufficient reason for essentially higher activities that exist in living things, except through the sort of substantial change described above. Such substantial changes, as found in the presently constituted natural order of things, take place solely by assimilation or generation.

The hylemorphist philosopher understands that things' substantial unity above the atomic level depends upon some unifying principle: substantial form. Only natural unities possessing appropriate cognitive faculties of sensation or intellection can know anything. A "sensing device," such as a television set running in an empty room, senses nothing. It cannot see its own picture or hear its own sound. No genuine perception occurs until, say, a dog stumbles into the room and glances at the running set. The dog can see and hear the set because the dog is a natural, living, substantial unity whose primary matter is specified and unified by a substantial form (its soul) that possesses sense faculties of sight and hearing. Absent the sensitive soul, the most complex sensing device knows nothing of the sense data it records. Absent the intellectual soul, a "thinking" machine understands nothing of the intelligible data it manipulates. A computer programmed to pronounce, "*Cogito ergo sum*," ("I think, therefore, I am") remains completely unaware of its own existence or anything else.

German mathematician Kurt Gödel unintentionally underlined electronic computers' inherent limitations. In 1930, he proposed his famed "incompleteness theorem" to the Vienna Academy of Sciences. In simple terms, Jaki tells us the theorem "states that even in the elementary parts of arithmetic there are propositions which cannot be proved or disproved in that system."[73] According to Jaki, Gödel's theorem struck "a fatal blow to Hilbert's great program to formalize the whole of mathematics."[74] Jaki maintains that it "cuts the ground under the efforts that view machines...as adequate models of the mind."[75]

Jaki spells out the incompleteness theorem's impact on the question of computer consciousness: For a machine to prove the unprovability of a system's formula is for it to be self-conscious. It would need to reflect on its knowledge, and reflect on that reflection, and so forth to infinity. A computer "would always need an extra part to reflect on its own performance."[76] But a machine "will not and cannot be fully self-conscious" because it will never "be able to reflect on its last sector of consciousness."[77]

Despite Jaki's adroit analysis, we must recall that machines possess no psychic faculties at all. They have no consciousness whatever.

Gödel's theorem implies that human beings are not machines. Computers cannot know the truth of their own judgments because they (1) lack a spiritual intellect and, thereby, (2) lack the capacity for self-reflective consciousness.

Computers may simulate language-trained apes' abilities. Incompleteness theorem analysis shows computer computations remain essentially inferior to human cognitive abilities. Apes and computers cannot perform genuinely self-reflective intellective acts. Only creatures with spiritual intellects, such as human beings, can perform such acts. Unlike computers, apes live and possess sensitive souls capable of sense consciousness, not intellection.

Electronic computers have no sensation or intellection, not even life. In principle, we can design and program them to imitate, or even exceed, language-trained apes' skills. These facts show that ape-language studies (1) pose no threat to our uniqueness as a species and (2) cast no doubt on our uniquely spiritual nature, distinguished from the rest of the animal kingdom.

Ape-language studies provide a historical incident that underscores radical difference between humans and lesser primates. No language-trained ape possesses greater reputation for linguistics expertise and presumed civility than the female chimpanzee, Washoe. Paul Bouissac reports an "attack of the celebrated 'Washoe' on Karl Pribram, in which the eminent psychologist lost a finger." He concludes that the incident "was undoubtedly triggered by a situation that was not perceived in the same manner by the chimpanzee and her human keepers and mentors."[78]

In pointing to the divergence of perspective between human and ape, Bouissac might understate the problem. Washoe would have been about 15 years old when she attacked Pribram. We have virtually no record of outstanding human students of that age attempting to bite their teachers.

Animals, even language-trained apes, do not perceive the communicative context the same way that human beings do. Such divergence of perception reveals the degree to which the anthropomorphic fallacy overtakes many researchers, despite their claims of caution.

The preceding discussion of human uniqueness focused upon signs of our spiritual nature and, to a greater degree, failure of lower animals to demonstrate intellectual abilities. Austin M. Woodbury offers an even more decisive perspective.[79] He argues that efforts to explain animal behavior in terms of sensation alone (1) could never be complete and (2) might produce only probable conclusions because of the task's complexity. Disproving anecdotal data could be endless.[80] To avoid this negative approach's logical weakness, Woodbury proposes direct and positive proof that brutes lack signs of intelligence's necessary effects.[81]

Woodbury maintains intellect's necessary formal effects are: (1) speech, (2) progress, (3) knowledge of relations, and (4) knowledge of immaterial objects. Since each is a necessary effect, "if it be shown that even one of these signs of intellect is lacking to 'brutes,' then it is positively proved that 'brutes' are devoid of intellect."[82] Woodbury argues that brute animals default in all four areas.

The most significant ape-language experiments were conducted after Woodbury wrote his *Psychology*. Our research confirms that his claim of absence of true speech among brute animals was correct. He points out that some animals possess the organs of voice (or, the hands to make signs), the appropriate sensible images, and the inclination to manifest their psychic states. Since they lack intellect, they manifest no true speech.[83]

Woodbury holds that, if brute animals possessed intellect, they would have long ago developed communication expressed in arbitrary or conventional signs. Their failure to do so reveals absence of intellect. Since human beings possess intellect, all healthy humans develop speech.

Chimpanzees brought up in a human family learn no speech. Human children generally do so easily and quickly. Granted, chimpanzees and other apes lack human vocal dexterity. Still, they possess sufficient vocal equipment to enable them to make limited attempts at speech, just as a human suffering from a severe speech defect. Apes attempt nothing of the sort.

Woodbury makes no reference to signing apes. Imitation and association of images explain their behavior. Humans artificially impose signing upon animals. Brutes' failure to develop language in their natural habitat on their own demonstrates lack of true speech. That animals possess natural signs is irrelevant.

And animals make no genuine progress. Woodbury maintains that "from intellect by natural necessity follows progress in works, knowledges and sciences, arts and virtue."[84] Animals learn from experience, imitation, and training. Because they lack intellectual self-reflection, they cannot correct themselves, an ability absolutely essential to true progress.

Human beings progress as individuals even in the most primitive societies. Children learn language, arts, complex tribal organization, complex legal systems, and religious rites.[85] Woodbury says, "Moreover, the lowest of such peoples can be raised by education to very high culture."[86] He argues that appetite to make deliberate progress inheres in beings endowed with intellect and will. As intellect naturally seeks universal truth, will seeks infinite good. No finite truth or good offers complete satisfaction.

Human beings, as a species and individually, continually seek self-correction and self-perfection. While apes are ever content to satisfy the same sensitive urges, we erect ever-advancing technology and culture: marks of civilization's progress. Animal failure to make non-accidental improvements, except when human intellect imposes them through training, proves absence of intellect within animal nature.

Woodbury claims brute animals lack the third sign of intellect, formal knowledge of relations. Animals fail to understand formal significance of the means-end relationship. While humans grasp the formal character of the cause-effect relationship in terms of being itself, animals only perceive and associate a succession of events.[87]

Woodbury distinguishes between (1) possessing universal understanding of the ontological nature of means in relation to ends as opposed to (2) possessing sensitive knowledge of related singular things. Lower animals reveal lack of such understanding when conditions change, making the ordinarily attained end of instinctive activity unobtainable. They lack ability to devise a substitute means to that end. They will repeat the now futile action that instinct presses upon them.

Woodbury offers the example of apes, accustomed to stand on a box to reach fruit. If we remove the box, they will place a sheet of paper on the ground and still try to reach the fruit.[88] This reveals how lower animals "show no knowledge of distinction between *causality* and *succession*."[89] Apes that understood causality would not conceive a sheet of paper as causally capable of lifting them significantly toward fruit.

The fourth and final sign that animals lack intellect pertains to knowledge of immaterial things. Our intellective nature impels us to knowledge of science, exercise of free choice, living a moral life, exercise of religion, and so forth.[90] Such abstract and supra-temporal objects so exceed the life of apes and other animals as to need no comment.

Brute animals, including apes, lack all four necessary formal effects of intellect: true speech, genuine progress, knowledge of relations, and knowledge of immaterial objects. Sufficient reason impels us to conclude that lower animals lack intellective faculties and, thereby, an intellective soul, since the presence of a given form necessarily implies its formal effects.

When we attempt to understand animal behavior, we need to make a critical distinction between (1) intellective knowing of universal concepts and (2) sensitive knowing through a common image or common scheme. Materialists incline to identify the two, thereby, confusing sense and intellect. Woodbury defines the common image or common scheme as an image of a singular thing according to sensible appearances that happens to be similar to other singular things.[91]

We can readily understand many modern animal researchers' errors because (1) image association grounds the entire sensitive life of apes and lower animals (including phenomena associated with signing behavior) and (2) we commonly confuse common images with universal concepts. These researchers suffer the same confusion as the eighteenth century sensist philosopher, David Hume, who conceived images as sharply focused mental impressions and ideas as simply pale and derivative images. Hume and the modern positivistic animal researchers misunderstand the essential distinction between image and concept.[92]

The distinction between image and concept manifests the radical difference between material and spiritual orders. Being rooted in the individuating, quantifying character of matter, the image is always singular, particular, sensible, concrete, and imaginable. We can easily imagine a single horse or even a group of horses. The concept is universal in nature, since it involves no intrinsic dependence upon matter at all. The universal (1) entails no sensible qualities whatever, (2) can have varying degrees of extension when predicated, and (3) is entirely unimaginable. No one can imagine "horseness." No single image of a horse or group of horses would fit equally all horses, even though the common image of "a horse" enables a fox to recognize sentiently all horses' sensible similarities. This "common image" is more useful for the instinctive life of animals. The common image of a mouse suffices for a cat's sensitive estimation that mice are fit objects to pounce upon and eat. Intellective understanding of a mouse's essence may obsess a professional biologist. Such understanding is hardly necessary, or even helpful, to a famished feline predator.[93]

Consider the following example designed to illuminate the difference between (1) sensitive recognition of a common image and (2) true intellective apprehension of an intelligible essence. Imagine a dog, an uneducated aborigine, and a civilized person all observing a train pulling into a station at the same time over successive days. All three would possess a train's common image. This permits sensible recognition of likeness of the singular things involved, sequentially observed trains. Whether these are the exact same engine, cars, and caboose is irrelevant, since similar sets of singular things could be known through a common image.

Sensible similarities are all the dog would perceive. The civilized person would understand the essence of the train. This would entail grasping inner workings of causal forces of fire on water producing steam whose expansion drives pistons to move wheels pulling the whole vehicle, cargo and passengers as well, forward in space through time. Well enough. What about the uneducated aborigine? What differentiates that "primitive" from the dog is that, even though the human person may not initially know the train's intrinsic nature, intellect is at once searching for an answer to the "why" of the entire prodigy. The aborigine might make what, to modern observers, would be amazing errors in this regard, as did the Borneo natives who offered animal feed to cargo planes landing there during World War II. But any healthy adult human intellect would search the existing causes of the train's inner structure and function. With some explanation, the aborigine would come to the modern understanding of the train, while the dog still would bark uselessly at its noise.

So, too, when human beings and mice perceive the same mousetrap, they perceive something quite different. The mouse sees the cheese. We see a potentially death-dealing trap. Witness the divergence of perspective between psychologist Pribram and chimpanzee Washoe concerning the proper role of Pribram's finger in the context of their communication. At every level of communication, animal perception is purely sensory; human perception is sensory and intellectual.

The mouse sees the cheese in a strictly sensory manner as the object of its totally sensitive appetite. We see sensitively in the analogous meaning of intellective "sight." The trap's deadliness is evident to us. From a past close call, a mouse might react in fear of a trap because it associates a trap's image with an image of earlier (non-fatal) pain. Human beings know why a mouse should be afraid.

Available animal studies are consistent with the above example and its explanation. This explanation better fits the facts, since brute animals, as Woodbury has shown, reveal that they lack the intellectual faculties we possess.

I have distinguished human beings from lower animals in two ways: (1) Presently available natural scientific evidence regarding lower animal behavior, including recent ape-language studies, constitutes no legitimate challenge to the essential superiority of the human intellect. (2) Woodbury's positive demonstrations prove the non-existence of intellect in lower animals. I have noted many unique capabilities and accomplishments of human beings, individually and collectively considered, that bespeak intellectual faculties utterly transcendent to the experiential world of brute animals. Human beings can understand natures, make judgments, reason, create true languages, make genuine progress, build great civilizations, worship their Creator, and even write treatises about their own nature and origins. Brutes cannot.

Six

THE HUMAN SOUL'S SPIRITUAL CHARACTER AND DIVINE ORIGIN

I do not intend my study to dwell upon topics whose full development belongs to the philosophical science of psychology. To complete this analysis of human nature, I will briefly consider the demonstration for the human soul's spirituality offered by St. Thomas Aquinas in his *Summa theologiae*.

In defending the intellective soul's spiritual nature, St. Thomas seeks to prove the soul is (1) immaterial, that is, not itself extended in space, and (2) subsistent, that is, it exists as a substance in its own right and is not in any way directly dependent upon matter for existence. He argues:

> The soul of man is an incorporeal and subsistent principle. For it is evident that man through his intellect is able to know the natures of all bodies. However, whatever is able to know certain things cannot have any of them in its own nature, since that which is in it naturally would impede knowledge of other things, as we see that an infirmed man's tongue which is tainted by a feverish and bitter humor is not able to perceive anything sweet, but rather everything seems to it bitter. Hence, if the intellectual principle had in itself the nature of any body, it would not be able to know all bodies. However, every body has some determinate nature. Thus, it is impossible that the intellectual principle be a body. And similarly, it is impossible that it understand through a corporeal organ because, if it should, the determinate nature of that corporeal organ would impede the knowledge of all bodies, just as if a certain determinate color were not only in the pupil of the eye but also in a glass vase, the liquid in the vase would seem to be of that same color.[1]

I will not try to analyze exhaustively St. Thomas's argument. Others have already done so.[2] Suffice the following: To understand the essence of all corporeal things, the intellect must itself be in potency to all such essences. It cannot already be any corporeal thing. In the knowing act, the knower becomes the thing known. If any particular body constituted the intellect's nature, the intellect would already actually be that thing. It would not be able to be simultaneously any other particular thing, much less all corporeal things. The intellect can know, that is, be, all corporeal things. Therefore, it cannot be any bodily thing itself.

Just as an eye's lens must be itself colorless so that we can see all colors through it, the faculty of sight must be in potency to all colors. Sight must be of no particular color itself. So, the intellect, whereby we know the essence of all material things, must be in potency to all material natures. No material nature can constitute it.

Someone might object that, while a television cathode ray tube has one physical form, it can image all visible forms on its physical surface. Even though the power of sight is intrinsically dependent upon the corporeal sight organs, it enables an animal to see all visible forms. Apparently, the intellect's receptivity of all corporeal forms does not demonstrate its spirituality.

On the contrary, a television tube's surface does not know, that is, become, objects represented on its surface. The cathode ray gun at the tube's back emits electrons which sweep the inside of the screen, thereby illuminating thousands of individual phosphors. The patterns of light and dark produced are discerned only by a knower with genuine sense faculties, for example, a dog.

Because the seeing faculty is itself neutral or potential to variant forms of color, it enables an animal to perceive color. Sense perception of color is different from intellectual understanding of a corporeal thing's essence. The power of sight is not a valid parallel here to the power of intellectual understanding. Sight's formal object is not the essence of things. The parallel is: just as sight cannot have color within its very nature intellect cannot possess a corporeal being's nature.

Because sense powers always attain their objects under material conditions, dependence of sight and other external and internal sense faculties upon material organs is evident. Through sensation, we perceive objects as extended in space and time, with sensible qualities, and so forth. The intellect's objects evidence no such dependence on matter. We apprehend them in a manner abstracted from such conditions.

Complete defense of this proof would exceed the scope of my investigation here. A second proof for the intellective soul's spiritual nature is taken from the strict immateriality of objects understood through intellection.[3] And a third proof is based upon intellectual self-reflection.[4]

The second demonstration rests on the intellect's ability to understand natures in a manner entirely stripped of material conditions.[5] Intellectual abstraction enables the intellect to form the universal concept. Aristotle and St. Thomas emphasize the mind's active role in producing its own act of understanding the essence of things. This demonstrates the intellective act's immateriality, since the act must be proportioned to its produced object, the concept.

The radically immaterial universal concept's nature reveals the spiritual nature of the intellect producing it. The uniquely spiritual character of human intellective knowledge is manifest when we reflect on the distinction between the universal concept and the material image described earlier. While human beings and animals share the sense world of images, we alone can form universal concepts in and through which we illumine the intelligible natures of all things.

The intellect's ability for proper and complete self-reflection shows the third proof of the human soul's spirituality. Reflection is proper when a power knows itself and not some other power.[6] It is complete when it knows its own nature.[7] We know our own thoughts and judgments. When we know that we possess truth about something we are aware of the conformity of our own judgment to reality itself. Our reflection is proper since it knows its own acts. In these same acts, we are aware of the objects and truths understood and of the nature of these acts of understanding and judging. Self-reflection is complete.

No material organ has such proper and complete self-reflection. No physical organ, not even the human brain, can return completely upon itself so as to compenetrate its being and physical extension. While one physical part can reflect another physical part, and that part another, and so forth to infinity, no part can apprehend itself, and no collection of parts can grasp their whole as a whole. That is precisely what we do by intellectual self-reflection.

Such proper and complete self-reflection must be the truly spiritual act of a truly spiritual soul.

Demonstration of the human intellectual soul's spiritual nature poses new questions: (1) Can the evolutionary process naturally produce a spiritual soul in manner similar to that hypothesized in the case of lower living forms? (2) If the spiritual soul cannot be so explained, what would constitute an adequate cause of its origin?

With respect to the origin of higher organic forms, John N. Deely argued that, as long as the totality of the interacting agents suffice to produce matter's requisite organization, the appropriate living substantial form will be educed without offending against causality. A spiritual form's origin is quite another matter. While matter's reorganization may account for eduction of organic forms, it cannot produce a human spiritual soul.

Writing in his *Summa theologiae*, Aquinas argues:

The eduction of act from the potency of matter is nothing other than that something comes to be in act which previously was in potency. But, because the rational soul does not have its existence dependent upon corporeal matter, but rather has subsistent existence, and because it exceeds the capacity of corporeal matter, as was said above, it is not educed from the potency of matter.[8]

St. Thomas had earlier shown that the spiritual soul "exceeds the capacity of corporeal matter."[9] This flows from the demonstration of the soul's spirituality. Since the human soul "does not have its existence dependent upon corporeal matter," its origin transcends what is sufficient to produce organic forms.

According to St. Thomas, the only causality adequate to produce the spiritual soul is what we term "creation." He argues:

> The rational soul is not able to be made except through creation, which is not true of other forms. The reason is that, since to be made is the way to existence, the way a thing is made must coincide with its mode of existence. For that is properly said to be which itself has existence, as it were, subsisting in its own existence. ...The rational soul is a subsistent form, as was maintained above. Whence it is expressly capable to be and to be made. And, because it is not able to be made from pre-existent matter, neither corporeal, because thus it would be of a corporeal nature, nor spiritual, because thus spiritual substances would be transmuted into one another, it is necessary to say that the rational soul does not come to be except through creation.[10]

St. Thomas argues by dichotomy. The spiritual soul is somehow produced from pre-existent being, or it is created. Creation means to cause being, presupposing no pre-existent being from which it is made.

Pre-existent being is corporeal or spiritual. Corporeal being educes only more corporeal being. Spiritual being exceeds corporeal beings' potentiality. And spiritual being cannot be transmuted into another spiritual being. Such transmutation requires form-matter (hylemorphic) composition excluded from a purely spiritual substance's nature. Without any material substrate principle to maintain continuity between prior and posterior states of change, change would not be change at all. Instead, change would be annihilation of the prior spiritual being and creation of its subsequent replacement. Such existential substitution does not fulfill the meaning of the term "change" because change entails some real intrinsic connection between before and after states.

St. Thomas concludes that the spiritual soul cannot come from any pre-existent being, corporeal or spiritual. The intellective soul must come to be by causation that excludes pre-existent being. It must be created.

Since St. Thomas has demonstrated that the spiritual soul is created, he next seeks to determine the adequate agent of such creation. He concludes that God must be this agent.

> Now God alone is able to create, for the first agent alone can act with nothing presupposed, since a secondary agent always presupposes something derived from the first agent, as was maintained above. However, what does something with something presupposed, acts by producing a change. And therefore no other agent acts except by producing a

change, whereas God alone acts by creating. And because the rational soul is not produced through the transmutation of any matter, thus it is not able to be produced except by God immediately.[11]

In his earlier-written *Summa contra gentiles*, St. Thomas similarly concludes that the human soul is immediately created by God because of its spiritual nature.

But since the human soul does not have matter as part of itself, it is not able to come to be from something as from matter. It remains, therefore, that it comes to be from nothing, and thus, is created. Since, therefore creation is the proper work of God, as was shown above (c. 21), it follows that the human soul is immediately created by God alone.[12]

Earlier in the same work, we find the previously demonstrated fact that creation is God's proper work.

Therefore, the proper cause of being as such is the first and universal agent, which is God. In truth, other agents are not causes of being as such, but causes of being this, as of being man or of being white. However, the act of existence as such is caused through creation, which presupposes nothing. For nothing is able to pre-exist which is outside being as such. On the contrary, through makings other than creation is produced *this* being or *such* being. For from pre-existent being is made *this* being or *such* being. Thus, creation is the proper act of God.[13]

Since no pre-existent being exists upon which the agent that creates may act, the effect's production requires infinite power. St. Thomas Aquinas argues this point in his *Summa theologiae*:

For if a greater power is required in the agent insofar as the potency is more remote from the act, it must be that the power of an agent which produces from no presupposed potency, such as a creating agent does, would be infinite; because there is no proportion between no potency and the potency presupposed by the power of a natural agent, just as there is no proportion between non-being and being.[14]

Elsewhere, I have explained the meaning of this text:

The principle which St. Thomas employs here is laid down when he says, "a greater power is required in the agent insofar as the potency is more remote from the act." For as power means the ability to produce being or to act, its measure is taken not merely from the effect produced but also from the proportion between what is presupposed by the agent in order to produce the effect and the effect produced. Thus, to make a chicken from pre-existing chickens requires a certain measure of power. But to produce a chicken from merely vegetative life would require even greater power; and to produce a chicken from non-living matter yet greater power. But to produce a chicken while presupposing no pre-existent matter at all clearly would require immeasurably greater power. It is immeasurable, as St. Thomas points out, precisely because "there is no proportion of non-being to being."[15]

The infinite power required for the human soul's creation resides only in the infinite being we call "God." God's infinite power and direct causality is the only adequate explanation for the human soul's origin.

Seven

THE QUESTION OF
EXTRATERRESTRIAL LIFE

Naturalism is the philosophical presupposition. Evolution its only adequate explanation of the facts. If no God, no possible supernatural or preter-natural intervention, exists, then the cosmos's experiential data demands an evolutionary rationale. The economy principle demands a common rationale for a world filled with anatomically and biochemically analogous organisms. Absent divine intervention, descent with modification must have occurred. Alleged factual difficulties with evolutionary explanations must be momentary distractions. The general theory is indubitable because no other explanation fits the facts. Evolution is unfalsifiable in principle.

This kind of reasoning was not always the case. Before Darwin, the same experiential data convinced scientists and theologians alike that Divine Intelligence had imposed on the world a common plan of creation producing analogous similarities in living things. Darwinian evolution offered the possibility of explanation without God. Once scientists became accustomed to a self-explanatory natural world, rational objections to Darwinism became irrelevant to evolutionary theory's continued acceptance. Naturalism was an engrained mental habit. And naturalism demands evolution in some form.

Given the philosophy of naturalism, only evolutionary processes can answer questions of human or other biological origins. At some time, human beings and other organisms did not exist. Since no supernatural origin is possible, some natural explanation is necessary. Life on planet Earth must be the product of a natural process. Before life was non-life. Life must arise from non-life. The role of the evolutionary scientist is to determine what natural process generates life, not to ask whether non-life can give rise to life.

No matter what objections or difficulties arise in experimentally verifying how non-life gave rise to life, the fact that non-life gave rise to life sometime in the past is never in question. "Whether" life so arose is not at issue, only "how."

If experimental investigation suggests that life could not have arisen on Earth, naturalism's evolutionary presupposition demands that life existed in outer space and came to Earth by some sort of seeding process. No creationist explanation is allowed. Since life is here, it had to come from somewhere. If not from Earth itself, then from the cosmos. If time here is too short, then we can draw upon cosmic evolution's eons. In every instance, evolution works because no other explanation is possible.

Since natural means generated life on Earth or elsewhere, in an appropriate physico-chemical environment, abiogenesis becomes necessary. Given the cosmos's vastness, such conditions must have occurred, universally and repeatedly. Life must permeate the universe. If life gave rise to sentient and intelligent organisms anywhere, then, given the statistical likelihood of environmental replication, life must give rise to sentient and intelligent organisms throughout the vast universe. Hence, the common theme of astronomers calculating the number of stars with planets, planets with life, life evolving to intelligent civilizations throughout the cosmos.

The logic is necessary. All must happen according to evolution's mandate. Creationist objections are brushed aside as momentary embarrassments in a long march of scientific progress. The inevitable outcome is a universe filled with E. T.'s. To the evolutionary scientist no alternative scenario is rationally intelligible.

The whole thesis is based upon the philosophy of naturalism. Nature is the only force allowed in the cosmos. Evolution becomes unfalsifiable in principle because it is based on naturalism's philosophy, not on natural science. Evolution theory is philosophy, atheistic naturalism, not science.

Earlier I presented some contemporary scientific evidence supporting a natural, extrinsic bio-teleological interpretation of the material universe. This would mean God orders cosmic fundamental chemical components toward emergence of living organisms of ever-greater complexity and perfection. This view of a universe naturally evolving toward living forms is entirely compatible with sound principles of natural philosophy and metaphysics. Assuming that fitting physico-chemical conditions pervade the universe, theoretical possibility exists that the cosmos may enfold uncountable numbers and unimaginable varieties of extraterrestrial species.

Hypothetically granting this scenario does not concede human life's natural emergence. The human intellectual soul requires God's special creative act. No purely natural evolutionary process can directly educe its spiritual form. Should the accidental effect of interacting agents be what Austin M. Woodbury calls matter's "ultimate disposition," no compulsion would exist for God to create the human substantial form (soul). Recall that this ultimate disposition is simultaneous with that form which actively determines its micro-organization. Form is always logically prior to matter, especially when the form is subsistent. While appropriate material organization may occasion a material form's eduction, it could never force a spiritual form's production (creation). For matter's human organization to occur God must freely create the spiritual soul.

Matter's accidentally produced micro-organization does not force the Divine Will to create the human soul to inform such matter. Instead, the Divine Will creates the human soul. In a specifically human way, the soul, in turn, simultaneously perfects the matter's micro-organization. Nature may prepare for matter's transformation into human life. Only God can effect the actual transformation. Organic forms, which depend upon matter for their existence, can be educed from matter's potentiality. Because they subsist, spiritual souls cannot be so educed.

One might argue that God reveals His intention to make multiple human populations throughout the galaxies. God designs lower creatures to be subordinate to human beings. Animal, mammal, and primate evolution set the stage for human emergence. Absent humans, such preparatory stages on other planets appear to contradict God's intention. God cannot reject his own intention.

God's intention might not be so easily determined. If planets can manifest God's creative glory without life, life can manifest His glory without human beings. God's intention to create human beings might not be evident until they are actually created. God is not forced to create the human soul through material evolution of lower organisms. Corporeal beings are ordered only to produce more corporeal beings. Woodbury points out that lower species are not essentially ordered to evolve into human beings. Every genuine natural species "is ordained to its own perpetuation, not to its own destruction."[1] Authentic transformism must produce its effects accidentally, not essentially.

One might argue that accidental causality, always implicit in the natural evolutionary process, presupposes superior effects arising from inferior agents' interaction. If God initiates the evolutionary schema, He must implicitly pre-ordain its end: emergence of rational animals. God would foresee causal agencies' natural interaction resulting in human evolution. In initiating this process, He thereby reveals His intention to supply, by special creation, the human souls requisite to complete intelligent life's evolutionary manifestation.

This argument presumes that the process, from complex organic molecules to simplest living cells to higher plants and animals to human beings, inevitably progresses to fulfillment throughout the cosmos, given its nearly inexhaustible plenitude of environmental opportunities. Whether the universe's physical structure warrants such optimism remains to be seen.

Many natural scientists today make statistical arguments claiming that proper conditions for life must permeate the universe. They assert that billions of planetary bodies suitable for life exist in our galaxy alone, with a certain percentage having advanced civilizations. Anthropologist John E. Pfeiffer claims, "It would be a miracle if life had arisen only on the planet earth, if evolution were not taking place in billions of solar systems."[2] He then anticipates planets in all developmental stages, from the barren to the pre-living to those with nuclear-technologied species to those whose species explore space.[3] Such conjectures gain wide public dissemination.

Sir Bernard Lovell and Sir Francis Graham-Smith, two influential astronomers who have been directors of Britain's Jodrell Bank Observatory, pose serious objections to such optimistic scenarios. They claim some fundamental weaknesses exist in standard theorizing about extraterrestrial life.[4]

These astronomers grasp the extraterrestrial-life argument. They see it leads to the conclusion that tens of billions of stars might possess planets suitable for life.[5] They acknowledge much circumstantial evidence based on irregularities of nearby stars' motion may reveal gravitational influence of planets upon them. Since their writing (1988), existence of planets orbiting other stars appears confirmed.

Graham-Smith and Lovell note recent speculation suggests interstellar clouds produce (1) a majority of stars surrounded with a nebula of gas and dust from which planets might evolve and (2) chemical constituents from which basic building blocks of life, such as amino acids and nucleotides, can form.[6] Both data favor belief that organic life may permeate the cosmos.

The only place in all physical creation where we know life certainly exists is Earth. Graham-Smith and Lovell observe, "The only available evidence about living organisms exists on Earth."[7] Any scenario of life's emergence on Earth is problematic. The fossil record reveals first living cells appearing within a relatively narrow cosmological time frame of less than one billion years after the first amino acids and nucleotides, the chemical building blocks of life, accumulated in our oceans.[8]

A billion years might seem sufficient for life's emergence from this primeval soup. Many natural scientists believe presumption of easy feasibility is popularized science mythology. In an earlier work, Lovell explains the problem: The claim that "chance and natural law" would produce simple organisms from organic molecules "is valid only if the probability of the right assembly of molecules occurring is finite within the time scale envisaged."[9]

Lovell maintains that assembling a small protein molecule with 100 amino acid residues would require some 10 to the 130th power trial assemblies to obtain the correct sequence. He estimates the probability of achieving this within the time frame allowed "is effectively zero."[10]

In their more recent joint work, Lovell and Graham-Smith describe the chances of getting the correct sequence of amino acids and nucleotides within Earth's actual time frame as "vanishingly small."[11] Lovell earlier reveals awareness of a stereo-chemical selection process, such as proposed by Sidney W. Fox. Lovell suggests such processes may help solve problems posed by pure chance combinations.[12] Yet, he and Graham-Smith do not think such a solution can explain emergence of life within the time available in Earth's early history. They maintain, "our knowledge of the transition from the non-living complex macromolecules in the early oceans to the primitive single-celled organisms is speculative and fragmentary."[13]

Graham-Smith and Lovell accept the views of many leading scientists, such as biologist Francis Crick and astronomer Fred Hoyle, that life did not arise on Earth.[14] These scientists think the probability of life arising within the time and space boundaries available on Earth are "so vanishingly small that the only correct scientific attitude to assume is that the transition did not occur on Earth."[15]

To avoid the vanishingly small chance of purely natural evolution from non-life to life taking place entirely on Earth, many scientists now conclude that life must have originated in outer space, only to be transferred to Earth during the time frame in which it appeared here. Graham-Smith and Lovell concede that most "conventional scientists have been very critical of this modern panspermia theory of the origin of life from space."[16] They maintain that no "conclusive arguments have yet been produced against this idea."[17]

Many scientists point to evidence for organic molecules in the outer planets and throughout the cosmos. They argue life could have arisen during the much greater time frame available in outer space. Mechanisms for transport across space have been suggested. Spores could hitch rides inside comets or asteroids, coming safely to early Earth. Spores have been known to dry out, remain stable for millions of years, and revive. Insects have been found in forty million-year-old amber, with viable spores in their stomachs. Panspermia has its defenders.

Another scenario points to similar problems with life originating on Earth. Life's evidence goes back some four billion years. Until some 3.8 to 3.9 billion years ago, Earth was too hot for life because of continued asteroid bombardment. This indicates impossibly rapid Earthly evolution, thereby demanding life origins in outer space. Atheistic materialists' only choice is to posit such explanations, since they cannot admit that God's direct intervention might have started life on Earth.

An alternative hypothesis can surmount the statistical barrier forcing these astronomers to conclude that life originated in outer space. Even vanishingly small chances can bear fruit if chemical arrangements occur through divine providence. I do not suggest God's direct preternatural intervention.

God's infinite power and providence entail that the cosmic creative act predetermine the ultimate interaction of all natural physical agents, even to the subatomic level. Questions posed by the confluence of free human agents and divine providence do not arise at the pre-life phase of cosmic evolution.

"Chance" or "odds" are names we give for our ignorance of the total effect of all the world's intersecting causal lines. Within that republic of natures that constitutes cosmic evolution's primordial phase, metaphysical "odds" do not exist, especially, from God's viewpoint as cosmic creator and designer.

To insist that vanishingly small odds cannot be overcome ignores some implications of the generally accepted Big Bang theory of cosmic origins. Lovell points out that the universe expands at virtually the exact rate needed to prevent its collapse. He maintains that, at the moment one second after its expansion started, had the rate of expansion been reduced "by only one part in a thousand billion, then the Universe would have collapsed after a few million years."[18] He contends, "the only universe that can exist, in the sense that it can be known, is simply the one which satisfies the narrow conditions necessary for the development of intelligent life."[19]

Lovell observes that this unique condition "produces the remarkable idea that there may not be a solution in the language of science."[20] Some philosophers understand this because they realize that God intends to create a cosmos in which intelligent life develops. That He does so in defiance of all odds is not remarkable to the theistic metaphysician!

To avoid the evident implications of such a well-designed universe, some thinkers suggest an oscillation model of the universe: this cosmos is one in an infinite series of recurrent fireballs that gravitational attraction's eventual victory causes to self-implode after distant eons. They argue that blind chance, that merely looks designed, might produce intelligent life. Intelligent life that can recognize its origins can occur only in universes that appear designed, since only universes chemically ordered to life can give rise to intelligent beings.

Graham-Smith and Lovell reject the oscillation model because they contend insufficient cosmic matter exists to prevent endless expansion.[21] Still, since the 1930s, other astronomers have speculated that an invisible physical reality called "dark matter" fills much of the cosmos. We cannot detect this matter, possibly composed of neutrinos, by optical or radio telescopes. It might constitute ninety percent of the total cosmos. If sufficient dark matter exists, the oscillation model might be validated. Recent calculations indicate that, even if 99 percent of the cosmos were dark matter, total mass would still be only one-tenth that needed to prevent endless expansion.

Even if sufficient matter exists, thermodynamics' second law poses another barrier to an infinitely oscillating universe. Increasing entropy assures that energy is lost as unreclaimable heat in each creation cycle. This guarantees a finite process. And the second law indicates that the early universe was highly ordered, not produced by pure chance.

In any case, eternal cosmic series of expansions and contractions would not be a paradigm of pure chance. They necessarily presuppose eternal givenness of universal, non-evolving, physical laws. Such givens include at least such presuppositions as universal material interactivity, some expansion dynamism, and gravity itself. These constitute universal and eternal order. Primordial chaos was never true chaos. As Jacques Maritain adroitly points out, chance always presupposes order.[22] Even an eternity of Big Bangs in no way escapes the need for a cosmic Intelligent Designer.

God is not bound by statistical laws. Though the statistician might consider chances of an essential evolution step vanishingly small, the divine plan can still encompass it. Divine preordination of physical factors leading to life is entirely consonant with vanishingly small probabilities. Lecomte du Noüy explains how slight chances permit, instead of prohibit, an event, such as life origins, in apparently impossible temporal brevity. He maintains that chances, however small, do exist. A rare configuration might occur at the start, not the end, of countless eons. It might take place twice, then never again. All this is fully "in accord with the calculation."[23]

Odds' calculations concerning life origins do not determine the matter's truth. God's omnipotent creative design can overcome mathematicians' probabilities. Recall Einstein's dictum that God does not play dice. Despite all contrary odds, if God eternally so chooses, life's totally natural Earthly origination is possible.

Such divine ordination is consonant with possible extraterrestrial life origins. Lovell and Graham-Smith suggest "that the transition from non-living to living material could have occurred only in the infinitely greater boundaries of time and space of the Universe."[24] They grant that abiogenesis may have occurred elsewhere in the cosmos, even though most scientists are skeptical that life on Earth came from outer space.[25]

Sidney W. Fox and other natural scientists insist that to base abiogenesis analysis on calculations of chance is misguided. Fox claims stereochemical selection deterministically explains life's origins, not once, but many times in Earth's early history. He points to synthesis of "proteinoid microspheres" as models of initial steps in life's emergence. He insists experiments producing them are so simple they "*have been repeated* by uncounted thousands of high school and college students."[26] Fox defends these proteinoids or "thermal proteins," recognized by chemical authorities, as "much like proteins," produced by heating amino acids, not by organisms.[27] Fox claims these experiments suggest "numerous repetitions of the same generative process on the primitive Earth, all alike due to molecular determinism."[28] Apparently, Graham-Smith and Lovell admit possible formation of proto-proteins from non-living molecules. However, they insist that no sufficient temporal window of opportunity existed on Earth for the transition from non-living molecules to living cells.[29]

Whether the transition occurred on Earth or in outer space, Fox maintains formation exceeded proto-proteins. He suggests that scientists have experimentally defined primitive cells or "protocells."[30] Supposedly, experimental scientists have already formed such protocells, just as they formed protonoids before them and experimental evidence exists that supports cells' origin from proteins.[31]

John W. Patterson says that scientists see "Prigogine's dissipative structures" to offer a "natural mechanism for self-organization, perhaps even for the genesis of life from nonliving matter (abiogenesis)."[32] He asserts that "imposing strong temperature, pressure, or composition gradients" on "laboratory-simulated, prebiotic broths" has produced structures with "remarkable similarity to the simplest known form of life."[33]

To support naturalism's claims that life arises from non-life, I have presented arguments from Fox and others favoring stereochemical determinism. This gives evolution its most plausible scenario for testing its philosophical acceptability. Still, the naturalistic scientific case might not prove entirely plausible.

J. W. G. Johnson attacks Fox's claims about synthesizing proteinoids. He claims that Fox uses dry amino acids, mixed in artificial ratios, under laboratory conditions never replicated on Earth. The products are "nothing like proteins."[34] They differ from proteins because they are short, floppy chains of mixed left-handed and right-handed amino acids, lacking the proper protein order.[35]

Biochemist Michael J. Behe tells us that scientists are "deeply skeptical of these experiments."[36] He recounts the history of origin-of-life experiments dating back to Stanley Miller's famed 1952 production of amino acids using electric discharge into an atmosphere of methane, ammonia, water vapor, and hydrogen.[37] After Fox's failure to produce real proteins, the 1980s saw scientist Thomas Cech propose the nucleic acid, RNA, as another starting point for naturalistic abiogenesis. Behe argues that protein production's "big problem is that each nucleotide 'building block' is itself built up from several components, and the processes that form the components are chemically incompatible."[38] He reports, "In private many scientists admit that science has no explanation for the beginning of life."[39]

Even granting artificial protein synthesis, claims about possible formation of living cells meet serious challenge. Walter T. Brown, Jr. former Chief of Science and Technology Studies at the Air War College, maintains no evidence exists "that there are any stable states between the assumed naturalistic formation of proteins and the formation of the first living cells."[40] He cites leading evolutionists admitting that they have no explanation of the first cell's origins.[41]

While I cannot examine every possible scientific objection against natural abiogenesis, one fundamental chemical difference between non-living and living things bears noting. Brown observes that laboratory-synthesized amino acids always form in equal amounts of mirror-image structures termed "left-handed" and "right-handed." He notes, "the amino acids that comprise the proteins found in living things, including plants, animals, bacteria, molds, and even viruses, are essentially all left-handed."[42] He maintains no way exists to produce just one form. And the "mathematical probability that chance processes could produce just one tiny protein molecule with only left-handed amino acids is virtually zero."[43]

My longtime friend and colleague, biochemist Wayne A. Gallagher made much the same point to me some two decades ago, wisely pointing out the possibility that someone might advance a reasonable explanation for this curious phenomenon. Biochemist Russell F. Doolittle advances the suggestion that "D- and L-amino acids are different structures in space; mirror images do not have equal probabilities of binding to surfaces."[44] Perhaps good reason exists why "the amino acids that comprise the proteins found in living things...are essentially all left-handed."[45] Brown also points to a similar rotatory selectivity occurring in living organisms' sugars. Such sugars are essentially all right-handed.[46] Doolittle's suggestion might provide similar possible explanation for this sugar phenomenon. Gallagher's prudent caution remains in order.

If no known natural process can replicate or explain abiogenesis, why do many scientists insist it occurred? I maintain that naturalism's philosophy demands that natural processes explain life's origin. If not on Earth, then in outer space. Conversely, if preternatural divine intervention is possible, then abiogenesis is not necessary. Scientific abiogenesis evidence remains more speculative than demonstrated. If no adequate theory for abiogenesis is yet known, present evidence offers no basis for belief that life exists in outer space.

A. E. Ringwood, Director of Australian National University's Research School of Earth Sciences, notes the difficulty facing creationist objectors to an evolutionary model. He maintains that, as with the solar system's nebula theory, insurmountable objections to a hypothesis frequently disappear over time.[47] He recalls physicists' early 1900s argument against Earth's great antiquity. They claimed our planet possessed too much geothermal energy to be so old. Natural radioactive heat sources' later discovery easily disposed of this once formidable difficulty. As time has passed, the synthetic interdisciplinary evolutionary model has done a remarkable job of overcoming many such objections.

Still, intellectual integrity restrains many scientists sympathetic to abiogenesis, forcing them to admit its limitations and assumptions. T. Encrenaz and J.-P. Bibring concede that "living matter, even in the most elementary form–i.e., being able to reproduce itself at the very least–has practically no chance of existing anywhere in the Solar System except on Earth itself."[48] Because of collisions between meteorites and primitive Earth ejecting rocks that "seeded" Mars and, possibly, other planets, astronomer Hugh Ross expects to find at least remains of microscopic life on Mars. Not vice versa. Such finds would not prove spontaneous generation of life on Earth or anywhere else.[49]

Encrenaz and Bibring claim that astronomers cannot now determine whether life occurs elsewhere in the cosmos. They say we have no present "idea of the probability of life arising from prebiotic molecules."[50] Perhaps these astronomers are not fully aware of stereochemical selection implications for abiogenesis of Fox's experiments. Or, they may be well aware of the weaknesses in such research. Perhaps communication fails on this topic between diverse scientific disciplines and, in some cases, even between scientists in the same discipline.

Such ambiguities underline the prudence of avoiding natural sciences' perinoetic investigations as a definitive basis for philosophical or theological judgments. The disturbing possibility that natural science may never resolve basic origin-of-life questions confronts philosophers and theologians. This epistemological barrier arises because of (1) the complexity of the multifaceted issues at stake, (2) the incompleteness of experimental evidence, and (3) the continual possibility of new and seemingly insurmountable objections. Until a complete step-wise process for producing a modern cell is reality, no one can reasonably rule out preternatural divine intervention needed to bridge some critical gap in the links leading to life.

Encrenaz and Bibring tell us astronomers see no reason why life should be confined to Earth, because the Sun is an ordinary star like "one hundred thousand million stars in our Galaxy alone."[51] This mindset adopts naturalism's presumption: Life's emergence after Earth's formation occurred without divine preternatural intervention. If life orbits our Sun, it must orbit countless others. If the naturalistic process occurs here, no reason precludes it occurring there. Until scientists experimentally demonstrate each abiogenesis step, this naturalistic schema remains scientifically incomplete.

This commonly proposed scenario ignores the possibility that conditions perfectly suited for abiogenesis exist only on Earth, or are achieved only by deliberate divine preordination overcoming vanishingly small odds. Nothing guarantees that any other planet in the entire cosmos necessarily duplicates conditions exactly required for life to appear.

Scientific materialism is the standard thesis's underlying presumption. The argument runs: Since no God exists to create it, life must have appeared here by chance or molecular determinism. Given the vastness of the cosmos, life must have appeared elsewhere as well. Instead, since God does exist, life could have been created exclusively on Earth by direct intervention or deliberate preordination. The materialist argument falls.

The evolutionist argues: Uncountable planetary numbers scattered throughout the cosmos must achieve some environments suitable for life. Still, no *a priori* proof can show that some required conditions are not found, against all odds, solely on Earth. Perhaps, some life-emergence mechanism was needed, one for which no scientific evidence exists to indicate whether it could have occurred on other planets. Pending discovery of life elsewhere in the cosmos, no method exists to demonstrate the non-existence of such a mechanism. Even stepwise biochemical production of life in our laboratories would not prove conditions elsewhere are conducive. Only actual discovery of life elsewhere in space would undermine this objection. Even then, we would need exclusion of some special divine preordination or intervention in that extraterrestrial context.

Finally, the typical scenario proving extraterrestrial life concludes a certain percentage of planets must contain intelligent life forms. While material evolution might account for sentient organisms' generation, without God's special creative intervention, it cannot produce a spiritual, intellective soul. No valid extrapolation from Earth's intelligent life extends to other celestial bodies.

The only way to be certain life exists on other planets is to find it there, especially in the case of intelligent life.

Philosophers confront the possibility that, in the foreseeable future, natural science might be unable to resolve the question of life on other planets. Even finding non-intelligent life in space does not reveal whether intelligent life exists. By common definition any intelligent organism is human. Even if such human life is out there, cosmic space's immense distances may preclude our ability ever to contact an extraterrestrial human being.

Material evolution alone cannot produce even a single spiritual soul. If human life exists elsewhere in space, God has specially created it, just as He did on Earth. God's ultimate aim in creating the cosmos might be to produce human beings. By itself human life on the Earth is sufficient to fulfill this intention. That rational animals exist on a multitude of planets is unnecessary. Greater quantity of planets inhabited with intelligent life forms adds nothing to the qualitative perfection already manifested by Earthly human life. Cosmic immensity, overwhelming to human imagination, bears witness only to God's infinite power. Once the divine intention to create human life has been achieved on Earth, the quality of God's actions on other planets is a non-issue.

Short of their actual encounter, no scientific evidence available presently or in the foreseeable future logically entails extraterrestrial life forms. Discovery of other planets covered with jungles teeming with every imaginable living species, including primates like ourselves in every way, save intellect, would not warrant our expectation to find extraterrestrial true human beings (rational animals). Since such spiritual-souled organisms originate only through God's free, deliberate creative act, we can never anticipate discovery of extraterrestrial intelligent life. We can only recognize that act after the fact, if it occurs at all.

Whether theological reasons preclude God's creating human souls on other planets merits investigation. I defer that judgment to competent theologians and the Church.

Eight

THE METAPHYSICAL STRUCTURE OF NATURAL SPECIES

The biological-species concept, based upon sensible accidents and reproductive isolation, differs from the classical philosophical natural-species concept, based upon essential properties. Biology would conclude that mouse and elephant belong to distinct biological species. Classical philosophy would consider them members of the same natural species, since they have the same vegetative and sensitive powers.

While the philosophical natural-species definition may satisfy the philosopher's intellect, it appears a scandal to common sense. Philosophers have the same ease as anyone else using biological classifications that conform to common experience. Metaphysically considered, no such biological species as "mouse" or "elephant" exists. Since both animals (as well other biological species cutting across several classes) manifest the same powers, only that unnamed species whose creatures possess all such vegetative and sentient powers exists. The term "animal" is not appropriate as a species name, since certain animals, such as oysters, do not have all five external senses. Oysters belong to a different natural species than mouse and elephant.

Metaphysically, what determines individual organisms to the same natural species is participation in the same substantial form, despite differing matter which that form actuates. That same substantial form is the ontological foundation for abstraction (simple apprehension): the mind stripping away matter's individuating conditions to produce the form's intellective expression, the universal concept.

An organism's nature realizes its formal sameness in potentially infinite individual expressions throughout time and space. That universally conceived nature stands in timeless contrast to the contemporary conception of continually evolving biological populations.

Conventional hylemorphic doctrine holds that all individuals' substantial forms in the same natural species remain fixed and stable despite differences in material organization. We may raise a difficulty. St. Thomas Aquinas maintains that potency is always proportioned to its corresponding act: "Potency however, since it is receptive to act, must be proportioned to that act."[1]

Matter is related to form as potency is to act. Organisms' material diversity must reflect corresponding substantial form diversity. St. Thomas states that "matter must be proportionate to form."[2] Reginald Garrigou-Lagrange notes that "act, in so far as it is a perfection, is limited only by potency, which is itself a capacity for perfection."[3] From this, we could argue that substantial form's intensity and remission must vary from individual to individual in the same species. Or, that the same individual's form would vary with natural material changes occurring over time. Such interpretation allows for "evolution" of substantial forms, gradually changing with evolving material dispositions of biological populations.

This argument is spurious. The hylemorphic proportion fallacy presented above is an imaginative impulse conceiving substantial form as quantitatively proportionate to receptive matter. But form, as form, does not vary in intensity. Instead, form is indirectly subject to quantitative extension or magnitude.[4] As received in matter, form enables substance to be the subject of quantity, through extension in space and through individuals' multiplicity in a species.

A given natural species' substantial form is perfectly proportioned to, or fitted to, a wide range of material dispositions. Note the greatly varied biological populations sharing the same natural species. Since a substantial form's specificity expresses its acts through the powers it exercises, a unique set of material conditions does not limit its essential actuation. Instead, substantial form is present and its powers act in and through a multiplicity of material organs.

These material organs exhibit analogously similar functions in widely diverse biological populations. As long as these bodies' material organization allows the same operations of specific powers, the same natural species' substantial form is present. Once a given organism exists, its substantial form continues to exist and inform its matter. A wide range of changes in matter's disposition occurs during life. Still, form persists and life continues until matter becomes so disorganized that substance no longer retains basic unity: its act of living existence.

This broad range of material organization suitable to a substantial form's actuation appears opposed by Austin M. Woodbury's attack on the intermediate material organization needed for interspecific evolution. Woodbury speaks of "micro-organization found in the germ cell" as strictly essential to an intermediary type needed for evolution between natural species.[5] Such exacting specificity of material organization needed for natural species differentiation appears radically opposed to allowing broad ranges of material organization within the same natural species.

Still, recall that Woodbury allows for evolution within natural species. He once defined natural species as identical with biological classes, a very broad species concept. We must understand his micro-organization of matter, determinative of natural species differences, in terms of the presence or absence of powers proper to a given nature. While such material organization is critical to the existence or non-existence of natural species' powers, this does not prevent, in principle, a wide range of such micro-organization fitted to those powers specific to a given species. All sighted species must have some type of material micro-organization proper to vision, but that micro-organization's expression could vary greatly (1) between biological species, (2) within biological species, and (3) in a single individual during its lifetime.

Matter's micro-organization proper to a given natural species is dynamic, not static. It varies throughout life and between individuals of the same natural species. A human corpse's macro-organization may appear more suited to human life than that of a human zygote, but its micro-organization is not. At all stages of a given natural species' individual's life, its matter's micro-organization is fitted to existence and operation. Whenever material disposition sufficiently disorganizes, life ceases. The difference between living and non-living material organization may be intra-cellular and perinoetically undetectable, but it determines natural species.

Substantial form permits matter's variation like the limited flexibility shown by a stretched rubber band. The band can expand easily over a wide range of extension, up to the point at which resistance becomes hard. Movement beyond that point results in immediate breaking. So, too, a given substantial form may exhibit great variation in material dispositions. Still, if the limit is exceeded, substantial change necessarily and simultaneously occurs. This allows wide flexibility of a substantial form's material conditions, within the individual and between same natural species' individuals. John N. Deely refers to the formal cause or organization as "establishing a 'reaction range' outside of which the organism cannot be pushed without ceasing to be itself."[6]

Substantial forms are fixed and permanent. Evolution among forms determining natural species' natures cannot occur. An organism cannot be "on its way" to being human. It is human or not. St. Thomas confirms this when he insists that forms do not admit of degrees of more or less:

That by which something has its species must remain fixed and constant in something indivisible. Hence, in two ways it may happen that a form cannot be participated according to more or less. First, because the thing participating has its species according to that form. And it is for that reason that no substantial form is participated according to more and less.[7]

And, again, St. Thomas affirms the same truth:

For the substantial existence of each thing consists in something indivisible, and every addition and subtraction varies the species, as in numbers, as is said in Book 8 of the *Metaphysics*. Whence it is impossible that any substantial form receive more or less.[8]

As noted earlier, the presence or absence of certain essential properties or powers determines every natural species. In turn, substantial form determines such essential properties. St. Thomas argues that "no substantial form is participated according to more and less" because "every addition and subtraction" of an essential power, such as rationality or sight, would vary the species itself. In this sense, "the substantial existence of each thing consists in something indivisible." Thus, no intermediate exists between rational and irrational animal. Species divide through the presence or absence of given specific essential properties or powers.

A crucial point. St. Thomas explains why substantial form cannot "vary in intensity and remission":

Now Simplicius assigns the cause of this diversity to the fact that substance according to itself is not able to receive more and less because it is being *per se*. And therefore every form which is participated substantially in a subject does not vary in intensity and remission. Whence, in the genus of substance, nothing is said according to more and less.[9]

One human being cannot be more or less human than another. Or, more to my point, an irrational animal cannot gradually evolve into a rational animal. An organism is human or not. No intermediate stage exists.

Should a human being, through misfortune, lose a limb, that person is not less human. In quantitative dimensions and function, a little less of a person exists to be human. Still, what is is fully human. While less material extension of spatial mass expresses the Pygmy's human nature, the diminutive Pygmy is equally as human as the towering Watutsi.

Differences in material expression of the same substantial form account for accidental distinctions marking diverse biological populations belonging to the same natural species. Every vegetative organism, from microscopic algae to giant sequoia, participates the same substantial form. While we might have difficulty imagining that the sensible differences between such organisms are purely accidental, merely matter's diverse organization, imagining the human zygote developing into an octogenarian, informed throughout life by the same substantial form, or soul, is equally troublesome.

Possession or lack of rational understanding's power determines membership in the human natural species. Intellectual understanding's slightest glimmer is infinitely superior in qualitative intensity to the richest sentient cognition. We understand, or we do not. In this sense, intellect's possession does not admit of more or less.

Still, some human beings are more intelligent than others. They manifest perfection of intellectual operations in greater degree than those less fortunate. Intelligence is a formal human quality that admits of more or less. Diversity in intelligence arises from concrete material conditions in which that form and its powers must act, not from a diversity in substantial form and its powers.

St. Thomas acknowledges diversity of habitual perfection when he allows that, if substance exists in matter, "that is, according to material dispositions, more and less are found in substance."[10] He grants material condition's diversity causes diversity of habits and dispositions among diverse individuals of the same natural species. Such variation in a habitual perfection's intensity and remission occurs

> according to participation by a subject, namely, according as equal science or health is received more in one thing than in another, according to a diverse aptitude or from nature or from custom. For no habit or disposition gives species to a subject, nor again do they essentially entail indivisibility.[11]

Individual natures, dispositions, and habits allow room for diversity of participation in powers' perfection from one individual to another in the same natural species. The animal kingdom's great diversity of material organization permits widely diverse perfections in various sense powers' expression. Hawks see more perfectly than bats, while bats hear more perfectly than hawks. Still, both possess powers of sight and hearing.

While substantial forms and their powers as such do not admit of more or less, since they are either present or absent altogether, powers' actual operations entail degrees of participation in their habits' perfections, owing to material conditions. Every vegetative organism possesses nutrition, growth, and reproduction. But each plant exhibits greater or lesser perfection of these powers' operations, depending upon matter's peculiar organization. So, too, each human being is fully human. But exercise of human powers is the subject of habit or disposition. These admit of more or less perfection since they are participated by each human being in a manner proportioned to material conditions of existence. In this sense, act (form) is proportioned to its receptive potency (matter) in a manner in which both terms are affected: The formal quality manifested in the power's exercise is of greater or lesser perfection according to the intensity of its matter's disposition.

The number of natural philosophical species is as small as the number of biological populations is immense. Among organisms' natural species, human beings are unique and essentially superior. Natural species' substantial forms are fixed and permanent in nature, while widely diverse material dispositions permit powers' concrete expression to vary greatly in diverse biological populations. In short, biological evolution's theory may be compatible with fixed and discrete natural species, and natural mechanisms may give rise to higher organisms without violating fundamental principles of natural philosophy or metaphysics.

Nine

NATURAL SCIENCE AND THEOLOGY

Until now I have employed source data and analyses drawn solely from natural science and philosophy. Now I intend to re-examine human origins. Questions in this context impact upon fundamental religious beliefs. I will now deal with evolution theory's theological implications. I will try to determine whether the intersection of secular evolutionary science and sacred theology can bear positive fruit. The role of the Christian philosopher here is to test the rational compatibility of these diverse areas of human speculation.

Whenever right reason and scientific evidence appear to raise bold challenges to religious faith, Catholic intellectualism's great tradition offers calm assurance that true faith and right reason never genuinely conflict. As St. Thomas Aquinas points out, religious revelation's Author and human reason's Creator are one and the same God: Truth Himself. Vatican Council I concurs that "while faith is above reason, yet there can never be any real disagreement between faith and reason."[1] The same God infuses faith and puts reason's light into the human soul. Vatican I concludes by dogmatically declaring, "Therefore, We define that every assertion opposed to the enlightened truth of faith is entirely false."[2]

As a Christian philosopher, I begin my study of the human species' origin assured that no legitimate natural scientific evidence or correct human reasoning opposes authentic God-given revelation. Still, my task is fraught with difficulty. Natural scientific evidence and speculation concerning human origins are tenuous, tentative, evolving in interpretation, and often controversial. I acknowledge the delicacy entailed in correctly formulating and interpreting the content of revelation in this same area. My proposed explanations or solutions risk offending natural science and revelation.

To minimize these problems, I will first present an overview of current paleoanthropological claims about human evolutionary origins, mindful that this science suffers grave perinoetic intellective limitations: It must make judgments of events long shrouded in the distant past's hidden recesses. As a matter of principle, I will adopt the most orthodox and traditional religious doctrines in this area so that (1) subsequent speculative solutions might face the most acute test (in this way, I hope, at least, to avoid compromising revealed content merely to placate inflated claims of scientific certitude) and (2) I might help eliminate solutions of inadequate scientific and philosophical rigor.

Presently, I will discuss human origins using specified parameters and assumptions. Since the time that Dutch anatomist and paleontologist Eugene Dubois announced discovery of Java Man in the early 1890s, paleoanthropology has embraced the standard theory of human evolution. It traces *Homo sapiens* (human beings) back to early hominid forebears appearing some four million years ago and entering our own genus (*Homo*) somewhat over two million years ago. We can lay out no exact scenario of human development because experts often disagree among themselves. The scenario I present is a representative composite of opinions, many of which are controverted by others. The dates I give for hominid candidates are those commonly given, usually based upon radiometric measurements, and often range from tens of thousands to millions of years ago. Scientific creationists do not accept this method of dating. Many claim the world is only six to ten thousand years old and offer their own supporting evidence. Mainstream scientists ridicule such claims. The following discussion presupposes the framework of the standard theory of human evolution and standard radiometric dating ranges.

Contemporary anthropologists trace human origins back nearly four million years to the earliest hominid fossils having distinctly human-like anatomical characteristics: *Australopithecus afarensis*. Earlier origins from prior arboreal primate stock remain shrouded in mystery and speculation. Later Australopithecines, dating from three to slightly less than two million years ago, bear the respective designations of *africanus*, *robustus*, and *boisei*. Among the more recent genus, *Homo*, which begins about the time of *Australopithecus boisei*, somewhat over two million years ago, we find such respective designations as *habilis*, *erectus*, *sapiens* (archaic), *sapiens* (Neanderthal), *sapiens* (Cro-Magnon), and *sapiens* (modern).[3]

Paleoanthropologists' theories concerning emergence of human intelligence absorb them in speculation about anatomical evolution's staging. The belief that hands free from weight-bearing function are essential to more active environmental manipulation, subsequent brain development, and facilitating emergence of human intellective ability, causes paleoanthropologists to seek evidence of upright bipedal motion.

Ernst Mayr departs somewhat from this standard scenario, suggesting "the ability to make tools contributed far less to this selection pressure than did the need for an efficient system of communication, that is, speech."[4] He affirms the basic thesis that the "acquisition of upright posture and bipedal locomotion was the key event in the evolution of the hominid line (alas, entirely undocumented by fossil evidence)."[5] He claims this new posture partially freed forelimbs for other functions and permitted hands to manipulate "tools, such as rocks, sticks, and bones."[6]

Evolutionists insist that, although various hominid populations have diverse species' names, modern human beings' origin and development has been gradual, with no sudden demarcating leaps. Fossils once placed in different genera or species now fall into a single classification. As anthropologists find more specimens and fill in gaps, they become more convinced human evolution is a gradually changing continuum. Kenneth F. Weaver claims it "becomes increasingly difficult to draw lines that say, 'Here one species ends and another begins.'"[7] T. A. Goudge claims we cannot say exactly when or where *Homo sapiens* appeared. He maintains no "primordial Adam," no "individual" to whom the term "first *Homo sapiens*" applies, ever existed. Human evolution is "not a discrete event but a continuous process."[8]

Apparently, paleoanthropology's standard theory of human evolution challenges the philosophical conclusion that the human species is unique and essentially distinct from lower animals. Is this appearance grounded in reality? Philosophical certitudes demonstrated earlier cannot be contradicted by legitimate natural science.

Paleoanthropologists often tout the genus *Australopithecus* as the first primate to exhibit clearly human-like anatomical features. Still, it lacks characteristics proper to human beings. Australopithecines do not appear to have used stone tools or fire, or possessed speech. Such claims are reserved for later and, perhaps, derivative cousins of the genus, *Homo*.

Many anthropologists defend transition from Australopithecines to *Homo*. C. Loring Brace points to Richard Leakey's find, ER 1470, as providing "a gratifying picture of one of the steps by which *Australopithecus* became transformed into *Homo*."[9] Brace dates the true age of this East Turkana fossil at "between 1.6 and 1.8 million years."[10] He describes a time, some 1.5 million years ago, when both genera walked the Earth. He states both were erect-walking bipeds with no differences from the neck down. But he describes the Australopithecine as having an ape-sized brain with large teeth and jaws, while the other has a braincase nearly twice the size, but with molars only half as large.[11] The large-jawed, small-brained hominid is *Australopithecus boisei*, an evolutionary dead end. The larger-brained one is *Homo erectus*, from whom we are claimed to have evolved.

Brace concedes that primate instrument use prior to two million years ago is pure speculation. He guesses, "in the absence of any evidence," that weapons were like "the proverbial pointed stick."[12] He assigns first use of stone tools to about two million years ago. Brace admits that "these stone tools are a pretty unimpressive lot–pebbles the size of a human fist or smaller with a flake or two removed."[13] Louis Leakey found them in deposits at Olduvai Gorge in Tanzania in the 1930s. He called them "Oldowan tools." Leakey also believed he found fossils of a new species he called *Homo habilis* or "handy man." Brace maintains Leakey's label is premature. He points out, "No definitive description of this specimen has yet been published, nor has the required careful morphological and quantitative analysis yet been done."[14]

Whatever the case with Leakey's find, paleontologists assign obviously sophisticated Acheulean stone tools to *Homo erectus* by the Middle Pleistocene period, some three-quarters million years ago.[15] Brace suggests that deliberate tool-use evidence occurs from as early as 1.5 million years ago. He claims that evidence from the half-million-year time span reflected in Bed I and Bed II at Olduvai Gorge reveals "the record of the development of a tool-making animal from a gatherer and occasional scavenger to a deliberate big game hunter."[16] About 300,000 years ago, *Homo erectus* allegedly gave way to *Homo sapiens* (archaic).[17] Brace claims technological improvement burst forth about 100,000 years ago, setting the stage for agricultural society's quick emergence.[18] The Neanderthals flourished from about 125,000 to 32,000 years ago. The Cro-Magnons succeeded them, dating from 35,000 to 10,000 years ago. Modern human beings supplanted Cro-Magnons.

Since modern authorities no longer deny full humanity to the Neanderthals and their successors, I will focus upon *Homo erectus*. This biological species allegedly possessed "more advanced speech," knew how "to control fire and cook food," and made "far better stone tools and weapons."[19] If we accept these *Homo erectus* claims, then, with greater force, we should hold that *Homo sapiens* (archaic) belongs with later true humans. Based upon anatomical studies, Mount Sinai School of Medicine's Jeffrey Laitman maintains that "full command of articulate speech did not likely develop until perhaps 300,000 to 400,000 years ago."[20] Mayr points out such speculation's limits: "Speech does not fossilize and all we can say about the origin of language is pure conjecture."[21]

I have reprised briefly what conventional paleoanthropology says about human origins: human evolution's standard theory. It paints a picture of gradual emergence from continuous and contiguous hominid populations stretching back to ape-like creatures existing millions of years ago. We see that anthropologists believe human intellectual abilities' first physical proof appears some two million years ago, especially as tool-makers. Paleoanthropologists reject any clear demarcation line between early primates and modern human beings. They embrace a gradualistic continuum of evolutionary movement in which consciousness, self-reflection, and intellective activity slowly emerge. This scenario's religious-belief implications are evidently anti-literalist. Whether we can rationally reconcile them with orthodox Scriptural interpretation remains to be seen.

Ten

THE TRUTHS OF REVELATION

As a Christian philosopher, I need to determine those Christian beliefs that directly bear upon the human species' origin and initial condition. The first three chapters of the Book of *Genesis* are its Biblical foundation. The Catholic Church's teaching magisterium has clearly identified essential facts whose literal and historical meaning Catholics may not call into question. because they touch upon Christian fundamental teachings. The 1909 Biblical Commission affirms these facts include

> the creation of all things which was accomplished by God at the beginning of time, the special creation of man, the formation of the first woman from man, the unity of the human race, the original happiness of our first parents in a state of justice, integrity, and immortality, the divine command laid upon man to prove his obedience, the transgression of that divine command at the instigation of the devil under the form of a serpent, the fall of our first parents from their primitive state of innocence, and the promise of a future Redeemer.[1]

Not all of these doctrines touch directly upon evolutionary science's evidence. Sanctifying grace is not subject to empirical speculation. And evolutionary theory cannot confirm or falsify concrete historical acts of God or human beings, such as (1) the divine command to Adam and Eve, (2) the transgression and fall, or (3) the promise of a Redeemer. God's creation of the world in time concerns evolution's preconditions, not evolution as such. My study needs to consider: "the special creation of man, the formation of the first woman from man, the unity of the human race, the original happiness of our first parents in a state of justice, integrity, and immortality."[2]

The Fourth Lateran Council (1215) dogmatically confirms the Biblical Commission's teaching on "the special creation of man." It explains that, after creating out of nothing the spiritual, angelic world and the corporeal, visible universe, God "formed the creature man, who in a way belongs to both orders, as he is composed of spirit and body."[3] In affirming that "God called the first man into existence," Ludwig Ott distinguishes the divine act as being "creatio prima" for the soul, but "creatio secunda" for the body.[4]

In his encyclical, *Humani Generis* (*Concerning Some False Opinions Which Threaten to Undermine the Foundations of Catholic Doctrine*), Pope Pius XII provides the basis for Ott's interpretation. He says that we may speculate about bodily evolution, provided we leave unquestioned the human soul's direct, immediate creation.[5] Provided that God directly created the human spiritual soul, we may read the Biblical Commission's "special creation of man" to mean that the human body's material origin took place through some form of evolution.

The Biblical Commission insists on "the formation of the first woman from man." Cyril Vollert notes that it does not say "from the body of the first man" and is completely silent about the "rib."[6] He argues that the close relationship between Adam and Eve need not be based on bodily derivation since *Genesis*' inspired author "uses an obscure Hebrew word, '*sela*', which can mean rib, side, flank, etc."[7] Concerning *Genesis* 2: 22: "And the Lord God built the rib which He took from Adam into a woman," Ott concludes that "the saying is and remains mysterious."[8]

"The unity of the human race" implies the doctrine of monogenism, that Adam and Eve are the first parents of the entire human race. *Humani Generis* explicitly rejects the opposing doctrine, polygenism. Polygenism holds that Adam represents a number of first parents or that, after him, true human beings lived on Earth who were not his natural descendants.[9]

Ott maintains, "The teaching of the unity of the human race is not, indeed, a dogma, but it is a necessary presupposition of the dogma of Original Sin and Redemption."[10] We might call it an "indirect" dogma. This "teaching of the unity of the human race" follows, as *Humani Generis* points out, because Original Sin "is the result of a sin committed, in actual historical fact, by an individual man named Adam, and it is a quality native to all of us, only because it has been handed down by descent from him."[11] The Council of Trent's Decree on Original Sin defines that Original Sin "is communicated to all men by propagation not by imitation."[12]

"The unity of the human race" means the entire human race takes its origin from Adam and Eve, our first parents, who were an actually existing individual pair of human beings, male and female, from whom we are all descended through natural generation.

The Biblical Commission also affirms "the original happiness of our first parents in a state of justice, integrity, and immortality." The "state of justice" means being in the state of sanctifying grace. Vollert tells us that an immediate effect of sanctifying grace in Adam and Eve was "easy control over passion."[13] He notes that "integrity," the lack of conflict between the sense cravings and noble spiritual desires, is indicated when Scripture says "they were both naked yet had no shame."[14] Vollert also notes that, through their disobedience, Adam and Eve lost bodily immortality and introduced death to the world.[15]

St. Thomas Aquinas explains the preternatural gift of integrity accompanying the supernatural gift of sanctifying grace in Adam and Eve:

> But the very rectitude of the first state seems to require that he was
> formed in grace.... For this rectitude was according to this: that his rea-

son was subject to God, the inferior powers to reason, and the body to the soul. However, the first subjection was the cause of the second and the third, since while reason remained subject to God, the inferior powers remained subject to reason, as Augustine says. It is manifest, however, that that subjection of body to soul, and of the inferior powers to reason, was not from nature–or otherwise it would have remained after sin, since even in the demons the natural gifts remained after sin, as Dionysius says in Chapter 4 of his *De Divinis Nominibus*.[16]

"Rectitude," the state of integrity, is subjection of a human being's reason to God so as to cause lower powers to be subject to reason and body to be subject to soul. In light of this proper internal ordering of Adam and Eve's entire being, St. Augustine maintains that integrity gave them the possibility of avoiding sin without difficulty.[17] Augustine also notes that the attendant preternatural gift of immortality should be understood as the possibility of not dying, rather than as being the impossibility of dying.[18]

The gift of immortality does not contradict every physical substance's potential corruptibility, its composition of matter and form. The possible immortality of our first parents in no way opposes the dictum, "Every man is mortal." This famed syllogistic axiom merely states the potentially corruptible character of human hylemorphic nature. Ott maintains that the associated gift of impassibility means "the possibility of remaining free from suffering."[19]

Is the standard theory of human evolution, as proposed by most contemporary paleoanthropologists, compatible with doctrinal historical truths concerning human origins as proposed by the Church's teaching authority? Or, does the present reading of empirical data refute Biblical "mythology?" Is there an unexplored alternative scenario? These questions stand at the center of my study.

Evolutionary science's perspective is fundamentally divergent from revelation's. Evolutionist Teilhard de Chardin recognizes this distinction. He claims paleoanthropology leaves room for the insertion of divine intervention or revelation in human prehistory. He says the "problem of monogenism" appears to elude science as such. As for Adam and Eve, he writes, "At those depths of time when hominisation took place, the presence and the movements of a unique couple are positively ungraspable, unrevealable to our eyes at no matter what magnification."[20]

Evolutionary science sees the broad picture of human origins taking place over a time-frame measured in hundreds of thousands, or even millions, of years. It cannot focus on events affecting a single pair of humans at a given point in time. Raymond J. Nogar concurs, seeing evolutionary science as working with "evolving populations" and "long-term trends," "not the individual happenings of a single pair at a certain day of the month."[21]

Anthropological data and theories are so general that they cannot oppose particular facts about Adam and Eve, unless even the broad trends of such data are shown to oppose such particulars' possibility. How can we know with certitude precisely what early humans were doing, say 300,000 years ago, when we were not there to observe them? Speculation based upon present data can, at best, indicate the nature and activities of early humans, pointing to largely undefined populations and imprecise time periods. It cannot address with precision the conditions of existence of a single pair of humans at a particular, distant-past time. It cannot exclude, *a priori*, the possibility of miraculous divine intervention whose reality falls entirely outside the fossil record.

Favorably to evolution's case, the 1909 Biblical Commission rejects unreasonably literal, strictly fundamentalistic, interpretation of *Genesis'* first three chapters.[22] It warns us not to expect *Genesis* to offer precise scientific explanations of creation's natural order.[23] In his 1943 encyclical, *Divino Afflante Spiritu* (*On the Most Opportune Way to Promote Biblical Studies*), Pope Pius XII encourages efforts of scholars loyal to the Church's teaching authority to find satisfactory solutions to "difficult problems" of Scriptural interpretation. Such "solutions" should "accord with the doctrine of the Church," especially regarding "the inerrancy of Sacred Scripture," and yet, "at the same time satisfy the indubitable conclusion of profane sciences."[24]

With respect to evidence relating to human origins, evolutionary science constitutes less than "the indubitable conclusion of profane sciences." And Nogar's "strong preponderance of converging evidence" offers no epistemological certitude.

Of God's three special gifts to Adam and Eve in *Genesis*, original justice, or sanctifying grace, presents no problem from evolution's perspective. Grace is not empirically detectable. By possessing intellective souls, the first true human beings are grace's fitting subject. Aquinas affirms that the first "man was created in grace."[25] He maintains that in "the state of innocence man possessed in a certain sense all the virtues."[26]

Not grace, but the preternatural gifts of integrity and immortality seem to violate early humans' natural condition. We find it hard to imagine their state as superior to our own. We do not easily control our passions, nor do we possess bodily immortality.

Still, preternatural gifts do not establish a natural condition. The preternatural is that which is *praeter naturam*, beyond nature. Unless we share with David Hume an absolute bias against divine intervention's possibility in the natural order, we cannot rule out these preternatural gifts in prehistoric human beings. Such gifts would not be manifest in the fossil record. They would be impossible only if they somehow contradicted the human essence or its essential powers. The proper ordering of elements in human nature that constitutes the gift of integrity is that nature's perfection, not its contradiction. So, too, immortality, the possibility of not dying, entails the possibility of continued life. Life is the first act and perfection of any living nature, not its contradiction.

Vollert remarks, "Adam's preternatural gifts can readily be reconciled with the data of prehistory and ethnology."[27] Nogar says little more, except to note that these special gifts could be given only to a truly human nature, intelligent and free, whatever be its physical appearance.[28]

These preternatural gifts of integrity and immortality appear to be the natural perfection of human nature, except that we presently tend to be ruled by our lower passions and obviously suffer and die. If so, considered *a priori*, our lower passions should be subordinate to reason and our bodily constituency should seek continued life. Sin and death are perversions of the natural finality of our nature and its powers. Except for their universality in our experience, sin and death appear abnormal to human nature as such.

Viewed from the intrinsic finality of our nature and its powers, the preternatural gifts are not strange, alien, or repugnant to the natural condition of early humans. Human nature's fallen condition represents the greater enigma to the philosophical anthropologist.

Again, from the standpoint of intrinsic finality, the proof that such gifts were *praeter*-natural is the dismal fact that they have been lost. If they were natural properties flowing from human nature, we could never have lost them. The preternatural gifts of integrity and immortality are beyond human nature. They represent simply the ultimate natural perfection of human nature, awaiting eschatological realization. Predictably, the fossil record gives no evidence of such gifts. Still, revelation presents no intellectual scandal if it maintains our first parents possessed them.

Theologian John A. Hardon points to a 1941 address given by Pope Pius XII to the Pontifical Academy of Sciences (30 November 1941) that raises an issue about Adam's possible primate forebears.[29] That address allowed no "allegorical interpretation" regarding (1) "the essential superiority of man" over other animals because of "his spiritual soul," (2) "the derivation in some way of the first woman from the first man," and (3) "the impossibility that the immediate father or progenitor of man could have been other than a human being, that is, the impossibility that the first man could have been the son of an animal, generated by the latter in the proper sense of the term."[30] Hardon supports the third point citing Pius XII: "Only from a man can another man descend, whom he can call father and progenitor."[31]

While this might seem to proscribe human evolution from lower primates, Pius leaves open "other questions concerning the origin of man" including "the degree to which a lower species may have co-operated in the formulation of the first man, the way in which Eve was formed from Adam, and the age of the human race."[32]

Regarding the position of *Genesis* on human evolution, Hardon maintains "nothing is directly affirmed as to how the body of Adam was formed." He insists that the second creation text implies "that God acted in a special way when he brought the body of the first man into being."[33] Hardon affirms that, since evolution has been popularized, "theologians have come to agree that transformism, or the evolution of the first man's body from a lower species, is compatible with the faith."[34] They add the "provisos" that (1) the human soul was immediately created by God and (2) God exercised special providence over the formation of the human body, "so that the first man was not literally generated by a brute beast."[35] Since even the patristic age countenanced the possibility of the human body's evolutionary origin, evolution "properly understood" is not a modern notion.[36]

Significantly, Hardon notes that "the real irritant regarding the origin of man is not so much Monogenism or Polygenism. It is the mystery of original sin."[37] He puts his finger on the central fact of revelation: If Adam and Eve did not exist as individual human beings who sinned, thereby communicating their fallen nature to all their descendants, there is no need for redemption or a Redeemer. The entire theological order is at issue in Adam and Eve.

Eleven

LOCATING THE FIRST TRUE HUMANS: "CURRENT THEORY"

The 1909 Biblical Commission insists upon God's "special creation of man." God's direct creation of the human spiritual soul and exercise of special providence over the human body's formation avoids conflict with an evolutionary origin of the human body. Earlier I philosophically demonstrated that God creates each human soul.

I have also shown that the intellective soul does not evolve, no slow "emergence of consciousness." No intermediate exists between irrational and rational animals. If we accept the current theory of human evolution, at some specific point in primate evolution, a first true human being appeared. True human beings must possess, however weakly, powers of intellect and free will. Prior to human beings' appearance, all previous primates were simply highly developed animals. While these animals might possess sensitive souls having a full complement of external and internal senses, they would have no intellectual understanding or free choice of will.

Following theories proposed by conventional evolutionists and applying properly philosophical criteria, my task now is to determine the likely paleo-anthropological population for that first human appearance. I will reflect philosophically upon evidence usually interpreted by natural scientists alone. I do not intend to supplant legitimate observations and speculations of experts in various special sciences. Instead, I hope that philosophical criteria of intellective activity may illuminate questions usually treated in non-philosophical fashion. My earlier analysis of ape-language studies might provide helpful insights in this task.

If we accept Aristotle's distinction that human possession of rational faculties distinguishes human beings from other animals, our search for the first true humans amounts to searching for evidence of intellective activity. Since speech does not fossilize, we seek physical evidence indicative of deliberate tool-making activity and deliberately-controlled use of fire.

Anthropologists do not consider the Australopithecines a likely population in which true humans appeared. C. Loring Brace claims the only hominid found between "3.5 and 2 million years ago" was *Australopithecus*, none of which "would be recognized as having attained fully human status."[1]

Still, some researchers have suggested that *Australopithecus robustus* (and possibly other early hominids) used wooden and bone tools, even though only later, stone tools have survived as fossil evidence.[2] Well-known chimpanzee researcher Jane Goodall highlights such possible behavior's significance. She maintains "tool-using performances" marked a major step in pre-human evolution. She claims that "when, for the first time, an apelike creature made a tool to a 'regular and set pattern' he became, by definition, Man."[3]

Goodall anthropomorphically sees this humanity criterion as applying to her chimps. She claims for them ability to make tools to "a regular and set pattern." She also asserts that chimpanzees have an "advanced understanding of the relations between things," an obviously intellective ability! Goodall claims her chimps modify objects for particular purposes to "a regular and set pattern."[4]

Goodall cites her now-famous observations of chimpanzee "termite fishing." She relates how chimpanzees judiciously select grass stems or other objects, hold them in their mouths, and proceed to out-of-sight termite mounds, using the newly-obtained tool to catch termites which foolishly crawl onto them.[5] Goodall connects her recent discoveries to typical anthropological speculations about early hominid tool-making, claiming her 1960 observations were the first-ever "of chimpanzee tool *making* in the natural habitat. Our early hominid ancestors undoubtedly used twigs and sticks long before they made the first stone implement."[6] She refers in her research to speculations that *Australopithecus robustus* used wooden and bone tools prior to later hominids' claimed production of stone tools.

Goodall reports other researchers' observations describing more impressive chimpanzee behavior. They tell of chimps gathering figs by climbing highest branches of a lower tree to reach a taller fig tree, whose lowest branches were not quite within their reach. They would break off branches, strip them of leaves and twigs, and use them as tools to pull down higher fig branches, sometimes using a hooked branch to depress a limb while grasping for it with the other hand.[7]

During early 1970s media exposure, such accounts of chimpanzee tool-making skills evoked enthusiastic and widespread public reaction, as if no subhuman species had ever before made tools. Still, naturalists have long known about lower animals' tool-making abilities, even to "a regular and set pattern." Ernst Mayr observes that some anthropologists appear unaware of widespread animal tool use. He gives examples, including such "unlikely animals" as sea otters and a Galapagos finch routinely using rocks or sticks to obtain food, bower birds using paint brushes, spiders using throw nets, and wasps hardening soil over nest holes by pounding with pebbles.[8]

Even non-naturalists well know that birds make nests to hold their eggs. Were it not for our ready explanation of such behavior in terms of instinct, we might be tempted to describe such nest-making activity as "a regular and set pattern" of fashioning selected twigs, grass, ribbons, and so forth, into a tool cleverly designed to prevent eggs from dropping to the ground. Normally, we would not acclaim such universally-known behavior as evidence of bird intelligence, even if someone like Jane Goodall suddenly discovered it.

Apparently, researchers are not as prepared to accept anthropomorphic interpretation of bird behavior as they are of primates that look more like human beings.

Some naturalists attempt to distinguish non-human from human tool-making by contrasting tool utilization to tool making in a regular and set pattern. They fail. If "making" means changing objects' natural forms to adapt them to new function, even clearly unintelligent birds do this with elements that they bend and twist as they weave them into their standardized nests.

As I will show, we can explain such tool-making as purely sentient behavior, thereby nullifying any claim that such activity is distinctive to human beings.

Earlier I presented Austin M. Woodbury's positive demonstration of absence of intellect in brute animals. They, including apes, lack all four necessary formal effects of intellect: true speech, genuine progress, knowledge of relations, and knowledge of immaterial objects. Sufficient reason impels us to conclude that lower animals lack intellective faculties and, thereby, an intellective soul, since the presence of a given form necessarily implies its formal effects. Whatever hypotheses we might advance to explain tool-making abilities of subhuman species, including Goodall's chimpanzees, intellective activity is definitely not involved.

In recent ape-language studies researchers trained chimpanzees and other primates to perform their activities. So, too, environmental experiences in the wild, over hundreds, even thousands, of generations have trained animals Goodall and others studied. This fact, combined with previously examined evidence pertaining to animals' imaginative faculties (including ability to associate images) and estimative sense (by which animals know unlearned beneficial behavior), and their ability to imitate and develop habits, suffices to explain the phenomena that Goodall finds so remarkable. No need exists for recourse to deliberate intellectual acts.

With respect to the accounts of termite fishing, statistical necessity requires that, given sufficient time, some chimps would have acquired such behavior by sheer accident. Initial learning experience would have occurred the first time a chimp, while playing with a grass reed or other suitable object, poked it into a termite mound and termites happened to crawl onto it. The pleasure accompanying instinctive thrusting of this unexpected food into its mouth would establish beginnings of a stimulus-response pattern, thus begetting an incipient habit. That such behavior should spread throughout Goodall's chimpanzee population merely bears testimony to chimpanzees' renowned capacity for imitation.

Goodall attaches significance to the distance between the point the animal picks up the tool and the termite mound (sometimes 100 meters). She thereby reveals failure to grasp chimpanzee ability to associate images based upon previous experience. Similarly, selection of appropriate materials, even to casting aside previously chosen ones, bears testimony only to the estimative sense's presence and operation, just as birds will selectively sort through and choose materials suitable to nest-building. Birds are not well-known for their great intelligence, to wit, the expression, "bird-brain!"

Although more impressive, in analogous fashion, we can explain chimpanzee behavior seeking figs by stripping branches and reaching from lower limbs. Without deliberate intent, human beings pull leaves and twigs off branches. So, too, chimps have learned such behavior. Instinctive actions combined with successful random activity and appropriate image association enables fig tree environment to train chimps toward proper fig-gathering performance patterns. Opportune use of hands and hook-shaped branches are within range of such learned behavior patterns.

As long as we center the process upon immediate food-gathering rewards, its purely sentient character is manifest. Simultaneous absence of intellective activity requires such interpretation. We need only compare reported chimpanzee behavior to the clearly deliberate acts which human beings would manifest under such circumstances.

We might speculate that, had a human being discovered the art of stripping leaves and twigs from branches, a mail-order booklet entitled, "A Thousand and One Uses for Stripped Tree Branches," complete with plans for making everything from roof-thatching to fishing poles, would have flooded the nation within a month. The human intellect grasps the universality of the means-end relationship. This understanding alone constitutes the essence of deliberate tool-making. Compared to this uniquely intellective activity, the particular means-end relationship learned through purely sentient experience bears weak resemblance. Chimpanzees will never go into the mail-order business because they have no intellect with which to understand the potentially universal character of the tools they have stumbled upon by chance. We can devise multiple and diverse tool applications only when we universally conceive a tool's essential efficacy.

Chimpanzee advocates, such as Goodall, would insist that we could interpret the equivocal character of some animals' behavior as intellective. Why, they might ask, do we insist upon a reductionist reading? The answer to such query is twofold: (1) We can read the data in terms of purely sentient activity. (2) We must so read the data because of Woodbury's positive proof that brute animals lack intellect. Goodall's conclusions contradict Woodbury's. Nothing in her field data can defend against the metaphysical necessity of his definitive analysis. The metaphysical principle trumps her conclusions.

The preceding section shows how many natural scientists misinterpret the essential nature of subhuman primates. The tendency to overstate research subjects' cognitive abilities, committing the anthropomorphic fallacy, arises from lack of proper philosophical formation and an excessively simplistic application of evolutionary theory to the problem of human origins. These scientists wrongly presume that, if humans are simply highly developed animals, then animals, especially other primates, must be underdeveloped humans. They uncritically accept gradualistic human intellective emergence. Goodall readily attributes understanding to her chimpanzees. Since she is convinced such human intellective acts occur in them, she also readily accords similar intellective activity to our "early hominid ancestors" whom she claims "undoubtedly used twigs and sticks long before they made the first stone implement."[9]

I must evaluate available paleoanthropological data philosophically, treating with caution conclusions of professional anthropologists who frequently do not know proper criteria for detecting authentic intellective activity. Impressive sentient activity of chimpanzees, apes, orangutans, gorillas, and other present-day subhuman primates may still be quite inferior to that of substantially larger-brained hominids which were, according to the standard theory of human evolution, in direct ancestry to and immediately preceded the first true human beings. Philosophically, I am especially concerned with actions that display evidence of internal senses: the imagination and estimative sense, and the human intellect.

Cognitive activities exhibited by our supposedly immediate predecessors would closely approximate genuinely intellective ones. My concern is to determine whether first, unequivocal, paleoanthropological evidence of strictly intellective activity exists. If discovered, this alone would assure that a certain hominid population would first evince true human beings.

I can easily show absence of Woodbury's four formal effects of intellect in present-day primates. Conversely, ascertaining whether such effects were absent in some prehistoric hominid population might prove forever impossible. A curious possibility arises: While anthropologists might not be able logically to affirm true humanity of a certain hominid population, philosophers might not be able logically to deny true humanity of that same population. We might be able to explain fossil record activities evidenced in purely sentient terms. Unequivocal formal effects of true intellect might also be present, but remain unrecorded in the fossil record.

Animals lacking intellect, such as Goodall's chimpanzees, can readily fashion and use tools of limited sophistication. Hypothesized use of wooden and bone tools by early hominids, such as *Australopithecus robustus*, provides no evidence that they must have possessed intellective activity. Even early production of stone tools usually associated with *Homo habilis* need not carry such inference. C. Loring Brace rightly describes these crude implements as "a pretty unimpressive lot–pebbles the size of a human fist or smaller with a flake or two removed."[10] Still, Kenneth F. Weaver claims that such Oldowan tools bespeak activities not found in modern apes. He notes that Bed I accumulation contains stones apparently taken from several kilometers away. He argues that "while modern apes pick up and use rocks as hammers, they do not try to carry them great distances or to sharpen them."[11]

While several kilometers is certainly much farther than the 100 meters that Goodall claims her chimpanzees carried grass stems and other material to the termite mounds, the difference is one of degree, not kind. In both cases, we can explain the behavior in terms of sense memory and association of images.

That modern apes do not sharpen rocks, and that *Homo habilis* did, might be true. This would not prove that intellect is required to perform this act. In its crudest forms, sharpening constitutes mere tool-fashioning, a skill that non-intelligent animals could acquire in much the same sensory manner as chimps learn to termite-fish.

Increasingly, respectable evidence suggests that *Homo habilis* might have been erroneously assigned to the genus, *Homo*. Instead of being an immediate ancestor to *Homo erectus*, as is commonly maintained, *Homo habilis* might have been a branching of the Australopithecines, entirely outside the main stem of the genus *Homo*. More recent finds by Richard Leakey, Mary Leakey, Donald Johanson, and others, support this conclusion. They indicate "that creatures anatomically more like *Homo erectus* existed at least at the same time as, and probably even earlier than did the original gracile australopithecines from Olduvai and Sterkfontein."[12] Anthropologist Charles E. Oxnard maintains, "The genus *Homo* may, in fact, be so ancient as to parallel entirely the genus *Australopithecus*, thus denying the latter a direct place in the human lineage."[13]

Conventional wisdom, which suggests a gradual transition from the Australopithecines through *Homo habilis* into the genus *Homo*, may be completely wrong. Oxnard's analysis suggests, "It is far more likely that the genus *Homo* is much older than currently believed and that the australopithecines of Olduvai and Sterkfontein represent only parallel evolutionary remnants."[14]

Production of stone tools that unquestionably reflect deliberate intellective activity is primary fossil evidence of human intellective activity. Australopithecines (including *Homo habilis*) and the actual progenitors of *Homo erectus* stretching back possibly several millions of years fail to provide such unequivocal evidence. No archeologically evinced tool-making activity by the above-mentioned primates exists which we cannot, in principle, explain in terms of totally sentient faculties and organs.

According to presently accepted theory, archeological evidence does not reveal Acheulean tools, usually associated with *Homo erectus*, until some 700,000 years ago.[15] These stone tools indicate true human intellective activity, evidence of the human intellectual soul.

C. Loring Brace refers to an "unbroken continuity of stone tools from the levels at the bottom of Bed I in Olduvai Gorge" up to the present.[16] Tool-fashioning alone does not decide presence of truly intellective activity. This activity must be the unquestionable product of true intellection. No one knows the exact date at which such tools are first evident. We might never be able to determine this with any exactitude.

Because some design aspects appear to lack utility, not because of their design's utility, Acheulean stone tools of the Middle Pleistocene apparently demand human intellective activity. They manifest aesthetic design elements. These tools are chipped in convincingly symmetrical fashion, not simply on one side or end, as seems typical of the Oldowan tools from Bed I. Acheulean tools appear to manifest artistic concerns. They are more than functional, rooted in possibly sentient singularity of particular means-end relationships. Goodall's chimpanzees' tools express functionality, fashioned through purely sentient capabilities for termite-fishing.

Acheulean toolmakers finish their tools with aesthetic concern to perfect the shape on all sides. They work, and are visually pleasing. Since grasp of artistic symmetry presupposes universal understanding of a geometrical ideal as a good concretely to realize, such tools reflect true intellective activity.

The creatures who made these tools were true human beings, not apes with active imaginations. Tools, that Brace graphically depicts as found in Middle Pleistocene deposits, are evidently of true human origin.[17]

As a philosopher, I am not directly concerned to determine the exact moment the human natural species first appears. I am concerned to ascertain criteria appropriate to such determination, especially because many natural scientists tend to err in this regard owing to unconscious, but mistaken, philosophical presuppositions. This estimation requires precise delineation between what is possible to the purely sentient soul and what must necessarily be attributed to intellect alone. Apparently, positivistic natural scientists are singularly unaware of this distinction.

Far less is written about fire as hominid intellective evidence than about stone tool fossils. Anthropologists Kathy Schick and Nick Toth indicate a possible explanation for this. They comment, "The earliest evidence of the hominid use of fire is controversial." They note a "major problem" is to find archeological data "that can serve as certain evidence of deliberate fires."[18]

A primary difficulty is the ephemeral character of fire evidence, especially as found under primitive conditions. Some suggestive data appears as early as 1.5 million years ago, but is difficult to analyze and open to equivocal interpretation.

Contemporary researchers have cast into doubt earlier claims concerning fire use by *Homo erectus*. Weaver has reported regarding *Homo erectus*, "According to evidence of charcoal at Zhoukoudian and elsewhere, he now learns to control fire and cook food."[19] About the same data, Schick and Toth observe, "At Zhoukoudian the evidence has recently been thrown into question."[20] They note a suggestion that ash-deposit layers, instead of being residual cooking-fire evidence, might simply be spontaneous combustion products of guano or other organic material that engulfed bones left by hominids or other animals. Claims of deliberate fire-use here might prove quite deceptive.

Schick and Toth conclude that "there is fairly widespread evidence for controlled use of fire by human groups" by the Middle Paleolithic era, but that, prior to one-hundred and fifty thousand years ago, "this evidence is controversial."[21] Thus, credible evidence concerning deliberate fire-use appears immaterial to determining the origin-time of the first true humans because (1) the first unequivocal evidence of human intellective activity appears in the form of artistically-designed Acheulean stone tools and (2) these tools may date to as early as 700,000 years ago.

The above-examined anthropological data might scandalize common belief in human species stability. We see a scenario of bodily evolution in which size, certain facial features, build, and even cranial capacity have undergone considerable modification, even since the first true humans' advent. Still, we need only consider remarkable variations found among present-day races to realize that human nature's fullness can express itself in widely diverse physiques. Physical appearance is of little consequence. Cranial capacity alone is no determinant of presence or absence of intellectual faculties.

Anthropologist John E. Pfeiffer points out that the cranial capacity of *Homo erectus* overlaps that of modern man. He tells us that "*Homo erectus*, which includes Java man and Peking man" had a cranial capacity which "varied from about 775 to nearly 1,300 cubic centimeters, with an average of nearly 975 cubic centimeters."[22] The overlap of upper part of the range with the range for modern humans shows that "some members of *Homo erectus* had brains larger than many people living today."[23]

While Pygmies are quantitatively smaller in overall size, and have a cranial capacity far inferior to their Watutsi neighbors, they are no less human. Although cranial capacity might account for some differences in intelligence as measured by experimental psychological techniques, even a mentally retarded human being possesses an intellective soul. The intellective soul is present as long as, however weakly, the ability to form true concepts, make judgments, and reason is present.

While some minimal material organization must exist for the intellective soul to inform matter, we cannot determine that minimal level *a priori*. Since activity reveals nature, the maker of aesthetically-symmetrical Acheulean stone tools must possess that minimal material organization needed for an intellective soul's presence, even should that maker prove to possess a cranial capacity of only 900 cc. Hominids earlier than *Homo erectus* might have evinced activities (artistic or other) proper to an intellective soul, but the currently-accepted archeological record provides no basis for such claims. Absent such purported evidence, paleoanthropology offers a reasonable candidate for study effort to reconcile natural scientific and theological data on the first hominid population. Middle Pleistocene *Homo erectus* appears to meet this criterion, provided we: (1) adopt the current theory of human evolution and (2) attribute Acheulean stone tool making to that hominid population. Given the scientific community's present wide acceptance of these two provisos, I will provisionally analyze the problem of human origins in these terms. At a later point, I will examine a startling alternative scenario.

Even well prior to the Middle Pleistocene, *Homo erectus'* physical appearance might not have been so different from our own. Richard Leakey's group discovered a 12-year-old "boy's" remains dated back to 1.6 million years. They describe the Turkana "boy" as "five feet four inches tall," tall for any boy that age today. Were he properly dressed and wearing a cap to hide his "low forehead and beetle brow, he would probably go unnoticed in a crowd today."[24]

Physical appearance does not matter. True intellect's presence does. Only the intellective soul's infusion into properly disposed matter effects true intellect. Conversely, genuinely intellective acts, such as those exhibited by Acheulean stone toolmakers, demonstrate an intellective soul's presence. Such symmetrical tools evince development of matter properly disposed for reception of the first true human substantial forms.

Without placing undue reliance on the current theory of human evolution, in what follows, for the sake of determining philosophical and theological compatibility, I will assume that Adam and Eve belonged to the hominid population known as *Homo erectus*. I will philosophically examine the implications of this assumption.

Twelve

ADAM AND EVE'S ORIGIN

I have already presented the elements that essentially prove God specially created the human race's first members. I have shown that only God can create the intellective soul, and that, if God directly creates the human soul, no special theological problem exists for a Catholic philosopher. The Church admits that some form of evolution might account for the human body's material origin.

Cyril Vollert indicates that the 1909 Biblical Commission "mentions the special creation of man, not only of his soul."[1] Special creation of the whole human being presents a surmountable problem, since the moment God created and infused (in the selfsame act) the first intellective soul into properly disposed matter, God created the whole human being.

Vollert argues that "the spiritual soul would radically transform the pre-existing animal organism it began to animate."[2] The previously purely sentient organism's radical transformation into true human being would be evident only through archeological signs of intellective activity. Anthropologists may date *Homo erectus* back over 1.5 million years. Still, assuming this hominid population did envelop the first true humans, nothing guarantees the natural human species traces back that far. For the biological species, or better, *Homo erectus* population might have existed prior to the first intellective soul's infusion into a *Homo erectus* individual. Such "special creation of man" would be evinced solely by discovery of artifacts unequivocally bespeaking intellective activity, such as Acheulean tools dated some 700,000 years ago.

The special creation of true humans will be manifest by the first clear signs of true human behavior, not in some dramatic and singular fossil phenomenon. Raymond J. Nogar, following the standard theory of human evolution, maintains that such creation and infusion of the first human soul "would not appear as a violent, miraculous act of Divinity," but as a "new mode of adaptation involving intelligence and free will, which is exactly what is found in the record of prehistory."[3]

God's special creation of the first true human beings could thus be consistent with fossil evidence of the first true humans. Paleoanthropology may provide data revealing the time of human origins. Yet, only philosophical criteria can confirm actual presence of genuinely human intellective activity within that data.

The 1909 Biblical Commission implies the teaching of monogenism, that Adam and Eve, an actually existing pair of individual human beings, are the entire human race's first parents. This aspect of Catholic teaching contains some difficulty. Nogar notes that evolutionists speak of "evolving populations, not individuals."[4] Most evolutionists would say "that a population (evolving) *could* pass through a bottleneck of only one mating pair but it is *not likely*."[5] Nogar points to a probability favoring polygenism. The evolutionary scientist "would consider the evolution of a population more likely than evolution from two individuals."[6] Still, Nogar affirms that God's providence could have provided special circumstances enabling monogenism to overcome such adverse probabilities: "'Special' does not mean *miraculous*, for even if *Homo sapiens* arose through the bottleneck of a single pair (monogenetic origin) the process would have been quite *natural*."[7]

Vollert makes much the same point, saying, "From the scientific point of view, Adam is simply a hypothesis that is admissible but unverifiable."[8] Recall that even Teilhard de Chardin concedes that science cannot scrutinize the activities of an unique pair of humans at some distant point in time so as to exclude the possibility of monogenism.[9]

Most evolutionists tend to describe human origins in polygenetic terms. This reflects their naturalistic philosophy which excludes divine intervention's possibility, even if that intervention is a natural pre-ordination for true humans to arise, as Nogar suggests, through the bottleneck of a single pair of first parents.

The first human body's direct, immediate formation and the first human soul's direct creation, an absolutely literal (if such is possible) reading of *Genesis*, constitutes no metaphysical problem. St. Thomas Aquinas, and other Church Fathers and Doctors, give this interpretation.[10] Still, when we consider divine intention in relation to the standard theory of human evolution which enjoys present scientific opinion's support, what would possibly be served by creating the appearance, but not substance, of human bodily evolution?

I attempt here to reconcile biology's evolutionary theory with the historical Adam and Eve of *Genesis*. In so doing, greater challenges arise than the bottleneck through which the monogenetic first family must pass. Vollert describes the complex alternatives faced in wedding theological truth with evolutionary anthropology.

Vollert rejects the first alternative, "that God transformed an adult organism into man," as having "only a tenuous connection" to evolution.[11] Still, God alone creates the human spiritual soul, even during natural procreation.[12] When some primate's matter had evolved to proximate disposition to a rational soul's activation, God could at any moment infuse a human soul, thus radically transforming its body's material disposition to suit perfectly that human soul. Adam and Eve could have been brought into being, body and soul, as fully developed adults. Even if imperceptible, such transformation would change an animal body into a true human body the moment it occurred.

Vollert's second alternative, that "the change was effected in the embryonic state" also challenges the imagination, but not inordinately.[13] Highly evolved non-human primates could well protect and raise such human children as their own. Such hominids would be their "parents" analogously at best! After suitable raising, "the human boy and girl may have separated from the horde and set out to lead a free life."[14] These first humans and their descendants may have avoided interbreeding with the nearly identical non-human population either (1) by reason of special divine ordinance and guidance or (2) through natural repugnance to sexual congress with animals lacking intellects. Absent the intellect's vastly superior survival capabilities, extinction of non-human primates forced to compete in the same ecosystems with true humans is not difficult to envisage.

On either of Vollert's suppositions, God infusing the soul into proximately disposed matter constitutes that matter's radical transformation: as miraculous an act as that portrayed in Holy Writ. Whether we conceive the "slime of the earth" (*Genesis* 2: 7) as some primitive dust or clay or as highly evolved "pre-human" primate, the human soul's creation and its active disposition of matter to itself is an act which God alone can perform. I say this in response to many authors who insist upon the direct and immediate production of the body of Adam by God, without allowing that pre-human evolution may have directly prepared for production of that same body. Transformation of unformed slime into the first human's body might require greater power than is needed to do so from a fit hominid. Still, God alone is capable of producing such an effect. Properly understood, theistic evolution need not challenge direct divine intervention such as Scripture clearly indicates on this point.

The 1909 Biblical Commission's remaining decree presents a somewhat more difficult and mysterious enigma: "the formation of the first woman from man." This Church teaching further complicates our attempt to understand how monogenism may have taken place. Vollert has shown (1) the pertinent Biblical text is open to broad interpretation and (2) the 1909 Biblical Commission carefully avoids forcing its literal reading.[15] Vollert also describes varied symbolic interpretations of the text offered by many recent scholars, such as saying that (1) Eve answers so fully to Adam's desires, "that it seems as if God had extracted her from his heart," (2) God "made the woman according to the ideal image which the man had of her in his mind," or (3) God used Adam's body as a kind of "exemplar cause" for that of Eve.[16]

These speculations, based upon literary form theory, avoid difficulties raised by literal reading, such as the imperfection of Adam prior to Eve's creation and the apparent loss of one of Adam's ribs.[17] Nogar makes similar observations of modern exegetical symbolic interpretations: "'From man' may be taken in the sense of an image, pattern or likeness, the religious and moral meaning being that she is the *same nature as man.*"[18]

Figurative interpretations avoid practical objections posed by literal reading, making understandable contemporary exegetes' embrace of them. Such scholars may be prudent and factually correct in so doing. Still, such liberty of interpretation is not conceded in every quarter. Theologian Peter Damian Fehlner insists that we cannot concede evolutionary origin of Eve's body in the manner in which *Humani Generis* (1950) permits scholars to propose hypothetically for Adam. He argues that the 1909 Biblical Commission "requires the formation of the wife from the body of the first man."[19] Fehlner then cites *Genesis* 2: 22-23. The Douay-Rheims version reads:

And the Lord God built the rib which he took from Adam into a woman: and brought her to Adam. And Adam said: This now is bone of my bones, and flesh of my flesh; she shall be called woman, because she was taken out of man.

Even if Fehlner is correct in saying the Church requires assent to "the formation of the wife from the body of the first man," recall Vollert's point that the inspired author "uses an obscure Hebrew word, 'sela', which can mean rib, side, flank, etc."[20] The "literal" reading of *Genesis* 2: 22-23 may be expanded to mean that Eve's body was physically educed from some part of Adam's body.

Robert T. Francoeur describes as "grotesque" some recent attempts to sustain the literal sense of the Scriptural text. He recounts suggestions such as Adam and Eve being "twins," formed from a double cell or Adam marrying a subhuman wife before he received a rational soul.[21] The twinning scenario, despite Francoeur's distaste for it, offers several distinct advantages. It retains pre-human evolution in our first parents' origin while affirming Eve's physical derivation from Adam's body. And it entails, as I will soon show, foreshadowing a most profound theological doctrine: the dogma of Christ's virgin birth.

Embryonic infusion of the first human soul, Vollert's preferred alternative, affirms the standard evolutionary thesis. The final mutation required to dispose pre-human matter for the first human soul's reception achieves fruition in the union of sperm and ovum at the very moment of the soul's infusion. Mutations expressed in the sperm and/or ovum which are needed for final preparation of matter to receive the human form do not themselves constitute the matter's preparation. Only when sperm and ovum are actually united is matter fittingly and specifically disposed for the human soul. At that exact moment, the infused intellective soul acts upon such prepared matter, radically transforming it into the first actually existing human being, Adam.

Nor should we be disturbed that such radical transformation takes place, as Francoeur notes, in a subhuman womb. If the evolutionary linkage is not to be broken, Adam must have been conceived in some subhuman womb.

Vollert describes the stage at which the human soul's infusion occurs as "embryonic." Webster's dictionary defines an embryo as "a vertebrate at any stage of development prior to birth or hatching."[22] The term "zygotic" expresses greater scientific precision. "Zygotic" is the proper technical term to describe the organism at the moment when sperm and ovum actually unite to form a new living human being. The soul would most fittingly be infused at that first moment of zygotic existence. This avoids the archaic successive animation theory.

The twinning of our first parents would have to be monozygotic if the body of Eve is to be physically educed from Adam. In dyzygotic twinning, separate ova are fertilized by separate spermatozoa, producing fraternal twins. Fraternal twins are in no way educed from one another. Their genetic relationship is no closer than that of normal siblings.

Monozygotic twins are the same sex since they share the same genetic structure. How, then, could the embryonic female, Eve, have arisen from physical division of the embryonic male, Adam? Such is not possible by natural process. God's miraculous intervention is required to give Adam a female helpmate formed from his own flesh. God must employ supernatural means to alter the new twin's gender derived from Adam's embryonic flesh.

This hypothetical intervention has a significant parallel event in theological history. A similar "gender miracle" is recorded in Christ's virginal conception. Instead of human male tissue being miraculously transformed into female, Jesus Christ was conceived as male, yet with no human male parent to confer the needed "Y" chromosome.

The case of Eve is purely conjectural. The virginal conception of Christ is defined dogma.[23] Both events contain a natural element with the miraculous being superadded. Both entail a process of generation proceeding from a human organism, thereby conferring human nature upon its issue. From Adam's male body, Eve must be miraculously educed as female. From the Blessed Virgin's female body, the male body of her Divine Son must be miraculously educed.

No comparison seems appropriate between this hypothetical prehistoric generation of Eve through twinning and the singular Incarnation of the God-Man Himself. Still, if our first parents' advent involved a miraculous aspect analogous in nature, but reversed in gender-direction, to that of the virginal conception of Jesus Christ, some parallel appears.

Recall the entirely hypothetical character of these speculations. I am engaged in what constitutes retrospective calculation in its most precarious

form. Every scenario consistent with evolutionary theory has evident difficulties. The proposal just examined preserves the natural connection of
Adam to his pre-human forebears. It defends Eve's origin from Adam's
body. Granted, this view of our first parents' life seems far removed from the
pages of *Genesis*. I do not propose to resolve every problem posed by excessively literal Scriptural interpretation. That task is better left to those more
qualified to judge the import of literary forms involved. My primary concern
here has been to find some hypothetical vehicle for reconciling the main the-
ses of evolutionary thought with authoritative Church teaching about *Genesis*.

We were not present to record actual events. God may have created our
first parents in the most literal Biblical fashion, leaving the archeological and
paleontological record in such fashion as to confuse and humble the intellectually proud. I have proposed a possible synthesis of scientific and theological data. I hope to illuminate some aspects of the question or to prompt
others to correct my errors so as to reveal the truth about this important speculative matter.

Some researchers have claimed that recent scientific discoveries shed
new light on the origin of a first common mother of us all. I will try to determine whether this data and speculation illuminate the nature or origin of the
Eve of revelation.

Until now, I have examined natural human origins primarily in terms of
the science of paleoanthropology. For more than a century a common presumption was that this enquiry fell under no other natural discipline and that
most fossil outlines of human origins were established. Then, some "upstart"
geneticists, mainly centered at the University of California at Berkeley, proposed that the search for human beginnings bears greater fruit when done (1)
far from the fields of Asia or Africa and (2) in total ignorance of stones and
bones. These geneticists audaciously proclaimed that they determined the
time and place of origin of our first human ancestors through research done
right in Berkeley genetics laboratories!

In 1987, three geneticists associated with Berkeley, Rebecca L. Cann,
Mark Stoneking, and team leader Allan C. Wilson, published research pointing to a single common female ancestor for all presently living human beings. They argued that mitochondrial DNA (mtDNA) from 147 persons
could be shown to "stem from one woman who is postulated to have lived
about 200,000 years ago, probably in Africa."[24]

Mitochondrial DNA became an investigative vehicle for restriction map-
ping designed to trace common human origins and chronology because it accumulates mutations several times faster than nuclear DNA. Also, unlike
recombinant DNA, mtDNA is transmitted exclusively through maternal
inheritance.[25] This permits techniques that trace our common ancestry to its
geographic and temporal sources.

Prescinding from technical analysis of this methodology, consider the Berkeley study's salient conclusions: First, given mtDNA's strictly maternal transmission, geneticists can construct "a genealogy linking maternal lineages in modern human populations to a common ancestral female."[26] Various possible evolutionary trees so constructed when combined with the parsimony principle imply that the tree of minimum length points to Africa as "a likely source of the human mitochondrial gene pool."[27] These Berkeley geneticists postulate a single African great-grandmother for all living humans.

Second, by assuming that human mtDNA sequence divergence accumulates at a constant rate, and assuming a 2-4 percent rate per million years, this research team calculated "that the common ancestor of all surviving mtDNA types existed 140,000-290,000 years ago."[28]

Third, these Berkeley geneticists offer the controversial suggestions that (1) all presently living *Homo sapiens* arose through some sort of transient or prolonged African bottleneck and (2) the product of this restricted genesis replaced all more archaic forms of the genus, *Homo*, virtually without interbreeding. The researchers concluded: "Thus we propose that *Homo erectus* in Asia was replaced without much mixing with the invading *Homo sapiens* from Africa."[29] The Berkeley geneticists' scenario aroused many paleoanthropologists' ire, seeming to deny their sacrosanct polygenism theory.

The initial enthusiasm for, and eventual weakening or rejection of, the most controversial tenets of Berkeley speculation is documented by Michael H. Brown.[30] He points out that the scientist who originally promoted use of "Eve" as a nickname for our mitochondrial common mother was Allan C. Wilson, lead geneticist of the Berkeley group.[31] Despite later attempts to disavow this confusing appellation, Wilson sowed the seed for ongoing controversy.

Brown points out that the Berkeley "Eve," unlike the Biblical Eve, "was probably not the one and only mother of all subsequent humanity."[32] He maintains "the mitochondrial Eve was probably never the only woman on earth."[33] He claims thousands of other women may have been alive at the same time, but that their mtDNA line became extinct by failing to procreate any daughters.[34] The "common female ancestor" this analysis pointed to may be simply a woman living some indeterminate time after the true Eve whose only real claim to fame is that she happened to be lucky enough not to have her mtDNA extinguished in human history.

The mitochondrial clock, so crucial to estimating "Eve's" origin time, came under revisionist scrutiny. The mutation rate of mtDNA may be much slower than initially claimed. If so, the time period between our common female ancestor's appearance and the present must be much longer. If the revised mutation rate were "1 to 2 percent instead of 2 to 4 percent," the time frame for this common female ancestor could reach as far back as *Homo erectus*, using the current theory of human evolution.[35]

Brown cites other authorities who challenge reliability of the mtDNA mutation rate as a measuring technique. They insist the rate has slowed down in higher primates, especially in human beings.[36] Thrusting the mitochondrial "Eve" back into *Homo erectus* would undercut the thesis that she was an immediate progenitor of *Homo sapiens*. This, in turn, would undercut the Berkeley groups' major thesis that *Homo sapiens* replaced more archaic forms without interbreeding.

The Berkeley geneticists' findings fail to constitute a legitimate, radical revision of conventional paleoanthropologists' views on human origins. No clear, convincing evidence exists that the Biblical Eve did not appear at the same time as first intellective activity appeared. The present theory of human evolution places this some 700,000 years ago. The "corrected" rate of the mtDNA clock points in just such a direction. Revelation still requires that Adam and Eve be the first true humans on Earth, the first parents of all true humans who follow them.

Much of the Berkeley geneticists' efforts were aimed at portraying modern humans' emergence in terms of *Homo sapiens*' genesis in Africa some 200,000 years ago. By so doing, they have tended to identify true humans with humans virtually identical anatomically to ourselves. Recall that the philosopher is less concerned with what a thing looks like than how it acts, for activity reveals nature itself. The first true human beings, Adam and his controversial spouse, Eve, appeared on Earth whenever animals with an intellective, and therefore spiritual, soul first appeared. Philosophically, the essentially distinct natural human species is so defined.

Thirteen

GENESIS: 4,000 B.C. OR 1,000,000 B.C.?

What is Earth's true age? When did true human beings first appear? Scientific creationists argue for a young Earth: 6,000 to 10,000 years old. Evolutionists reflect scientific consensus: some 10 to 15 billion years since the Big Bang. According to the most recent estimate, a 4.6 billion year old Earth, genus *Homo* dating back somewhat over two million years. Biblical literalists measure *Genesis'* patriarchal genealogy at barely two millennia from Adam to Abraham. Standard human evolution theory suggests *Homo erectus* made Acheulean symmetrical stone tools, beginning about 700,000 years ago. If true, evidence of such intellective activity would indicate that true humans appear at least by that point in the fossil record.

If true humans were cast from Eden at least some 700,000 years ago, can we plausibly accept them awaiting the promised Redeemer's coming through tens of thousands of generations? Can Sacred Scripture's proper interpretation accommodate the radically disparate chronologies of young-Earth fundamentalists and of old-Earth evolutionists? The cultural dominance of scientific materialism and evolutionism among modern intellectuals reduces the *Genesis* story to mythology: tolerated as myth, ridiculed by rational science. No serious analysis of human origins can avoid responding to this temporal enigma.

Biblical fundamentalists and others claiming literal Scriptural interpretation date human origins to some six millennia ago. Genealogies in *Genesis* 5 and 11 reflect the patriarchy from Adam to Abraham. Cursory reading supports temporal brevity. The list begins with Adam, giving his age as 130 years when he "begot" Seth. Seth begot Enos at age 105 years. Enos begot Cainan at age 90 years, and so on, until Thare begot Abram (Abraham) at the age of 70. The age at which each patriarch begot his offspring is stated in every case. Simple addition calculates Adam's creation to Abraham's birth at slightly over 2,000 years. Scholars now acknowledge compelling evidence dating Abraham slightly less than two millennia before Christ. From Adam's creation to present time computes to some 6,000 years.

This literal Biblical reading falls short of contemporary natural scientists' chronology by a factor of at least one hundred! Must we choose between natural scientific evidence and Holy Scripture? No reconciliation appears possible. Still, Christian philosophers should fear no seeming inconsistency. We know that true religion and true science never contradict. Some explanation is in order.

In some instances, Biblical usage of the term "begot" denotes generation of immediate offspring: son or daughter. In *Genesis* 11: 27, Douay-Rheims version, we see that "Thare begot Abram. ...And Aran begot Lot." *Genesis* 11: 31 documents the immediate filiation entailed in these begettings: "And Thare took Abram, his son, and Lot the son of Aran."

Those instances wherein "begot" entails multiple intermediate generations upsets the literal interpretation of patriarchal genealogies. In *Matthew* 1: 8, we read: "And Joram begot Ozias." When we turn back to 2 *Paralipomenon* 21-26, we discover that Ozias is not Joram's son. Joram is Ochozias' father. Ochozias is Joas' father. Joas is Amasias' father. And finally, Amasias is Ozias' father. Four generations separate Joram from Ozias, despite Matthew's declaration that Joram "begot" Ozias.

The above texts support Catholic scholars, who, from early in the twentieth century, have maintained that we cannot prove the genealogies to be continuous. Nineteenth-century Protestant Biblical scholar William Henry Green (1825-1900) argued that *Genesis*' genealogies fail to support any chronological inferences.[1] In a carefully reasoned article, Green presents multiple Biblical examples showing: (1) Genealogies are frequently abbreviated by omitting less important names. The most striking case is *Matthew* 1: 1 which speaks of "Jesus Christ, the son of David, the son of Abraham."[2] (2) Individuals with variant relationships are sometimes listed together without clarifying comment. Thus the opening verses of *Chronicles* (1 *Paralipomenon* 1: 1-4) list a "line of descent from father to son," but conclude with the last three being brothers.[3] (3) Scripture offers no collateral information whatever "which can be brought into comparison with the *Genesis* 5 and 11 genealogies for the sake of testing their continuity and completeness."[4] (4) Although *Genesis* 5 and 11's genealogies list each patriarch's age of begetting, Scripture offers no summations of these chronologies to indicate the author's purpose.[5] (5) Moses' direct and immediate knowledge of Egyptian civilization's antiquity precluded his being unaware that "the interval between the deluge and the call of Abraham must have been greater than that yielded by the genealogy in *Genesis* 11," assuming a strict computational chronology.[6] And, (6) the structure of *Genesis* 5 and 11's genealogies (each group including ten names) intends to reflect symmetry, not completeness, just as the genealogy in *Matthew* 1 is adjusted "into three periods of fourteen generations...brought about by dropping the requisite number of names."[7]

Green concludes that (1) "the Scriptures furnish no data for a chronological computation prior to the life of Abraham" and (2) "the Mosaic records do not fix and were not intended to fix the precise date either of the Flood or of the creation of the world."[8]

Genesis' number of generations becomes indeterminate. Such interpretation renders conceivable any time span for human existence that natural science's persuasive evidence might require. Though difficult to imagine, even millions of years' duration might prove consistent with Holy Scripture's genuine inspiration and inerrancy in *Genesis*' genealogies.

As for God leaving human beings awaiting the promised Redeemer for thousands of generations, recall that each individual in any waiting period is in the same theological situation. Even the literal *Genesis* reading leaves those cast from Eden in the same theological position for four thousand years as those immersed in a million-year or greater time span. Each such human being would begin and end life awaiting the Redeemer's coming. The longer period is no worse for the individual person than the shorter.

When Christ descends into hell to release the souls of the just to enter heaven, the quantity involved does not affect the status of any individual soul. The greater the number of souls awakened from the dead to enter eternal bliss, the greater the cumulative joy in heaven!

Whether short or long, the time span from Adam and Eve's fall until Christ's coming appears unproblematic.

While scientific creationists and evolutionists debate radically different estimates of cosmic age, philosophers should approach such arguments with caution, ever mindful of perinoetic disputations' epistemological limitations. To explore the extensive literature of these debates would exceed our enquiry's proper scope. Still, some comments are in order.

Acceptance of extreme cosmic antiquity is concomitant with the dominant evolutionary worldview. Long past the days of merely counting tree rings, contemporary evolutionists rely heavily on radiometric dating to validate gradualistic evolutionary time scales measured into millions of years. These techniques include potassium-argon, thorium-lead, rubidium-strontium, samarium-neodymium, lutetium-hafnium, rhenium-osmium, and two uranium-lead methods. Although nearly universally accepted, these dating procedures have critics.

J. W. G. Johnson attacks radiometric dating methods because they assume (1) a closed system unaffected by extraneous introduction or loss of elements over millions of years, (2) radioactive decay rates which are never affected by variables, and (3) initial absence of the radioactive decay product being measured.[9] The fact that such factors can cause errors does not prove that radiometric dating is not effective on a statistical basis. Expert geochronologists conduct their analyses to minimize or eliminate the negative impact of possible aberrations.

Most experimental scientists widely use and defend radiometric dating. Still, many non-radiometric dating methods indicate great terrestrial and cosmic antiquity. Early in 1996, scientists at Russia's Vostok Station in Antarctica drilled the polar ice shield to a depth of 3,348 meters. The extracted ice core revealed annual layering indicating nearly 400,000 years time span. This non-radiometric dating method offers results problematic to young-Earth advocates.[10] Daniel E. Wonderly describes some seventeen major types of non-radiometric data suggesting lengthy chronologies.[11]

Walter T. Brown, Jr. notes that all dating techniques presume that presently observed processes have always occurred at known rates. He warns this presumption may be unfounded: "Projecting presently known processes far back in time is more likely to be in error than extrapolation over a much shorter time."[12]

This points to the unavoidable logical weakness of retrospective calculation. Modern experimental science dates only from the last few hundred years. Modern scientific observations of nature's processes and their stability are limited to that time frame. To presume, as every long-term dating method must, that past events can be calculated backward from presently observed processes risks error caused by unexpected past events. Observations made over several hundred years may be misleading when extrapolated backward for time spans of millions or even billions of years.

Still, natural physical processes, such as radioactive decay rates, appear to be stable over vast time periods. Perhaps always. If nature did not always tend to the same end, act in the same way, natural science would be a useless passion. Miracles could never occur. God could not suspend non-existent natural laws. Uniformity and universality of natural physical laws are essential to scientific understanding of physical nature. I do not suggest physical laws have varied in the past. I suggest retrospective calculations involving extended physical law extrapolations risk error caused by interfering past events that reset the clock.

Astronomers retrospectively calculate that Thuban, not Polaris, was the pole star in 3,000 B.C. Earth's pole has a slight circular "wobble" which causes a change in pole stars over several thousand years. During a 26,000-year cycle, the pole stars will sequentially be Polaris (the present one), Gamma in Cepheus, Vega, Thuban, and back to Polaris. Based on this knowledge, if you ask an astronomer what was the pole star 5,000 years ago, the answer (Thuban) is determined by mathematically backing up this polar procession to the appointed time and determining which star should have been over the North Pole at that time. Although many Egyptian temples constructed about 3,000 B.C. were designed so that the beams of Thuban shone on their altars, astronomers do not go back in time and ask Earth's inhabitants to which star its polar axis then pointed. Should any major cosmic event have occurred in the interim, such as close passage by another celestial body, the calculation may be in error. Polar procession, like radioactive decay, is a calculable natural process. Still, an extrinsic causal event can disturb its proper result. Retrospective calculation, especially respecting prehistorical events and objects, entails this weakness of ignoring prior potential causal interventions. It might affect every method of dating distant past events or objects.

Another method of demonstrating great cosmic age is based upon light's speed. At 186,300 miles per second, astronomers measure remote celestial objects' distances in light years. The nearest star, *Proxima Centauri*, is 4.3

light years from Earth, meaning that it takes more than four years for light to travel from its surface to Earth. Some galaxies and quasars appear extremely distant. According to the most recent estimate, scientists conclude the universe must be some 10 to 15 billion years old, since it would take that long for light to reach Earth from these galaxies. This argument assumes that the speed of light is constant. Physicist Stephen Hawking proclaims, "There is only one absolute. ...It is not time but the speed of light."[13] But what if the speed of light were not constant?

In the early 1980s, mathematician Trevor Norman and astronomer Barry Setterfield, both Australians, sparked heated debate by publishing claims of speed of light degeneration. Their more recent 1987 update offers data from 163 separate measurements of light's speed in dynamical time made in the last three centuries (in part, to answer critics claiming their earlier analysis omitted relevant data).[14]

Norman and Setterfield claim that, during the last three centuries, experimental observations of the speed of light have shown a small, but measurable, decline as measured in dynamical time. Scientists define dynamical time in terms of Earth's orbital period. They define atomic time in terms of an electron's orbital period about a hydrogen atom. The authors claim that a least squares linear fit of the 57 best speed of light (c) readings from 1740 to 1983 indicates a c decay rate of 2.79 kilometers per second per year with "a confidence of 99.99% in the decay correlation."[15] Engaging in their own retrospective calculation, these researchers argue that available historical evidence indicates a non-linear decay curve in the speed of light best expressed in terms of an exponentially damped sinusoid.[16] In practical terms, this means that as we go back in time the speed of light increases exponentially so that a point some 20 billion years ago, atomic time (following Setterfield's value for the age of the universe), would correlate with a point only some 5,800 years ago, dynamic time, with c itself being about 87 million times c now. One million years ago, atomic time, becomes only some 4,800 years ago, dynamic time, with c about 70,000 times c now. The Biblical-evolutionary implications of this claimed correlation are obvious.

Throughout my study I have used the standard atomic time scale of the current human and cosmic evolutionary worldview. If Norman and Setterfield are correct, these atomic-based dates differ greatly from dynamic reality. The first true humans, dating some 700,000 years ago in atomic time (based on claimed association with Acheulean symmetrical stone tools), actually lived only a few thousand years ago, measured in terms of Earth orbiting the Sun. Following their new time scale, the time frame of true humans' possible appearance within the population of *Homo erectus*, would correlate easily with Biblical Adam and Eve. All radiometric dating would have to be adjusted to dynamic time.

Controversy continues to surround Norman and Setterfield's research into light's speed. Even its base data appears in dispute, especially three-century-old light-speed readings that they claim are noticeably higher than today's value. Distinction exists between (1) the methodology used to determine whether light's speed has decreased in recent times and (2) the mathematical determination of the curve best fit to describe that alleged decrease in order to extrapolate back to the cosmos's origin. Both aspects have met serious challenge from the scientific community.

The most curious of all young-Earth arguments is based on God's ability to create when and how He chooses. J. W. G. Johnson vigorously argues the point.[17] If, as metaphysics demonstrates, God creates the cosmos, God can work miracles. If God can work miracles, He can create the universe as He chooses, giving it the appearance of age though it be but an instant old. At the wedding feast of Cana, Christ turned water into wine, wine apparently fermented through a natural aging process as every good wine, but only seconds old in reality.

Or take the case of Pierre de Rudder (1822-1898), the Belgian peasant whose leg was broken so badly in 1867 that "after the fragments of bone were removed, the two bones that remained intact could be seen in the wound over an inch apart."[18] The lower part of his leg swung about like a rag with the open wound abscessed and suppurating for more than eight years. Following prayer to Notre Dame de Lourdes at a Lourdes Grotto replica at Oostacker in 1875, he was instantly healed, without even a limp remaining. Careful medical examination revealed the astonishing fact: "the instantaneous growth of a piece of bone over an inch long, filling up the gap where it was wanting."[19] Surely that new piece of bone gave the appearance of normal growth and development occurring through aging, but it was instantaneously created.

Johnson also mentions the obvious case: Adam. God could have created him as an adult. If so, he would have all the bodily appearances of having grown up, a process of aging to maturity. Johnson suggests that, when Adam gazed skyward at night, "he would have seen God's handiwork in the stars of the sky without waiting years for light to cross space."[20] By the same token, Johnson maintains that a geologist testing a rock sample in the Garden of Eden "would have found some potassium and some argon and he would have dated the rock at millions of years–the newly created rock."[21]

While Walter T. Brown, Jr. accepts "creation with the appearance of age" as a sound concept, he thinks it unacceptable in the case of starlight since (1) when a supernova explodes, the light reaching Earth might represent a star that never existed since the rays would be created *en route* and (2) beams of light would contain emission spectra representing a star's actual surface, not cold, empty space in which the rays were actually created.[22]

Could God create the impression of a universe fixed up with the appearance of age for ten billion-trillion stars? God, who creates and sustains the existence of each and every creature down to its least subatomic part, can create the cosmos any way He wishes and with any appearance He wishes. But would He do so? Would this constitute deception on His part? What appears unlikely is possible to God. His ways are inscrutable. Nor ought He be accused of deception for creative miracles such as those described above. If miracles create "natural deceptions" to heal creatures or as signs of God's purpose and love, is it deceptive to create a natural world such as ours just prior to humans' own creation? Perhaps Johnson is correct in suggesting that God "produced a fully-fledged, perfectly operating, adult universe."[23]

Still, while natural deceptions might be needed to sustain individual miracles, the entire cosmos's universally deceptive creation is a different matter. When everything appears miraculous, miraculous suspension of natural physical laws loses all meaning. Such all-pervading misrepresentation in the universe's basic structure appears to violate the intended cosmic intelligibility essential to the God-given gift of natural science itself. The appearance of age argument appears unsound.

Should light-speed decay advocates prevail, young-Earth Biblical interpretations find support. Even if we assume the last three centuries' disputed light speed decay, extrapolating a decay curve backward twenty billion years on so little data represents retrospective calculation in the extreme, with corresponding negative impact on the probability of the theory's accuracy. If light speed varied downward since Roemer's calculation in 1675, how can we be sure it did not vary upward prior to that time? In any case, the theory's perinoetic claims lie outside the competence of natural philosophy. Should Setterfield and Norman's claims prove unfounded, the argument for extreme cosmic antiquity based on light travel-times from distant celestial bodies retains its present force.

Arguments based on perinoetic intellective knowledge cannot produce certitude. Those natural scientific analyses favoring extreme antiquity, such as radiometric dating, ice core drilling, and light travel-time calculations, and those casting doubt on, or opposing extreme age, such as radiometric dating's criticisms and speed of light decay theory, fall into this category. Retrospective calculations always entail logical weakness. Still, since such calculations can give rise to varying degrees of probability, conventional natural science regularly employs such imperfect logic with apparent practical success. My evaluation of these probabilities is that young-Earth arguments lack logical rigor because radiometric and non-radiometric methodologies converge to support long chronologies.

The initial argument defending *Genesis*' ability to embrace extensive time spans remains sound. Legitimate natural scientific data regarding human origins poses no rational threat to Bible history. Revealed teaching appears compatible with a *Genesis* locale ranging from 4,000 B.C. to 1,000,000 B.C. Perhaps, much longer if need be.

Fourteen

HUMAN EVOLUTION'S EXTINCTION?

I have offered speculation showing how we might reconcile human evolution's current theory with divinely revealed teaching on human origins. My effort might be in vain. Old evidence, newly rediscovered, suggests the standard scenario might never have happened. Epistemological and factual challenges to human evolution's presently accepted explanation might exist.

J. W. G. Johnson writes about a modern man's skull and the remains of a woman and two children found in Pliocene strata at Castenedolo, Italy. The woman's cranial capacity was 1,340 cubic centimeters, well within modern range. The Pliocene Period is from two to five million years ago, well beyond modern human remains' acceptable limits according to human evolution's presently accepted theory.[1] Walter T. Brown, Jr. cites earlier works by J. D. Whitney, Malcolm Bowden, Frank W. Cousins, Sir Arthur Keith, and others. He points to remains of "modern-looking humans" which "have been found deep in rocks that, according to evolution, were formed long before man began to evolve."[2] Evolutionists tend to ignore or dismiss as mis-dated such evidence of true humans existing long before the standard theory would permit. Since these fossils date back to, or even prior to, the time of purported modern human ancestors, such as *Homo habilis* or *Homo erectus*, acknowledging their validity would be disastrous for modern evolutionists. We cannot have evolved from hominids who lived with or came after us.

In 1993, a thoroughly scholarly, 900-page work appeared. Co-authored by Michael A. Cremo and Richard L. Thompson, *Forbidden Archeology* offers detailed analysis of all significant paleoanthropological research done in the last two centuries.[3] This work examines relevant literature pertaining to every major hominid claim or find and materials on other fossils and human remains that the evolutionary establishment has long overlooked or suppressed. Although this work has vigorous critics, Cremo and Thompson's methodical thoroughness and often good logic make many of their analyses and central inferences hard to trump and unscholarly to ignore. *Forbidden Archeology*, praised by Phillip E. Johnson, but condemned by Richard Leakey as "pure humbug," was bound to stir controversy.[4]

In a twenty-two page *Social Studies of Science* review, Jo Wodak and David Oldroyd offer at least guarded acknowledgment that Cremo and Thompson have contributed to paleoanthropological literature because (1) "much of the historical material they resurrect has not been scrutinized in such detail before" and (2) they "do raise a central problematic regarding the lack of certainty in scientific 'truth' claims."[5] Wodak and Oldroyd maintain that "lack of certainty" arises "because the status of all knowledge is inherently a matter of degrees of probability and emerges as the result of social negotiation in concert with observation and inference."[6] They suggest that "those scientists who insist that evolution is a fact might be better advised to recast this as 'highly probable theory.'"[7] I suggest that philosophical and theological science produce certitude. Evolution's probability status arises because of natural science's perinoetic knowledge limitations.

Cremo and Thompson are not evolutionary materialists or Biblical creationists. They openly state Hindu affiliation as Bhaktivedanta Institute members. Following Vedic literature, they hold that the human race is of great antiquity, hundreds of millions of years old. For this reason, many critics attack *Forbidden Archeology*, claiming its authors' belief system precludes unbiased handling of subject matter. Such personal attacks are unjust and ill founded. Every author has a philosophical stance which might, but need not, negate objectivity. *Forbidden Archeology's* historical evidence and argumentation stand on their own merits as sociological and epistemological critiques of contemporary paleoanthropology.

Earlier I discussed current human evolutionary theory's compatibility with divine revelation. My primary concern with *Forbidden Archeology* lies in analyzing the epistemological challenge it poses to present human evolutionary theory.

Two hundred years ago, biologists, archeologists, geologists, and anthropologists tended to be creationists, not evolutionists. They saw sedimentary fossils all over the world, especially on mountaintops, as proof of the Great Flood. Catastrophism engraved Earth's surface, not evolutionary gradualism. In 1859, Charles Darwin's *Origin of Species* turned the tables of intellectual respectability. Gradualistic transformism prevailed. Thomas Huxley, Ernst Haeckel, and Charles Lyell raised the issue of intermediate primates leading to human beings. In 1871, Darwin finally endorsed the "*Descent of Man*." The search was on for human evolution's "missing links."

According to Cremo and Thompson, in the last half of the nineteenth and early part of the twentieth centuries, paleoanthropologists found apparent evidence of anatomically modern humans in Pliocene and even Miocene Periods.[8] When Dutch anatomist and paleontologist Eugene Dubois brought forth Java Man in the early 1890s, the scientific establishment proposed and soon embraced a true Pleistocene missing link, intermediate between early primates and modern humans. Java Man became the foundation for human evolution's current theory that presently pervades our culture from highest academic circles to popularized television documentaries.

Assuming Dubois's Java Man was a genuine hominid, would this logically prove human evolution? Not if modern humans existed at the same time or in prior eras. Dubois's find was erected as human evolution's foundation stone. Henceforth, any and all comers must fit the Java Pleistocene timetable or suffer scholarly extinction.

Although the now-discredited Piltdown Hoax partly supported human evolution's standard theory, further discoveries, at such African locations as Olduvai Gorge, Hadar, Taung, and Swartkrans, ensconced this theory in academe. While paleoanthropologists before Java Man held little bias against evidence for anatomically modern human beings in Early Pleistocene, Pliocene, or even Miocene Periods, later evolutionists could no longer tolerate such anomalous finds.

Influenced over many decades by "a biased process of knowledge filtration," science forgot discoveries that supported modern human beings' existence at times earlier than the standard theory would allow.[9] Cremo and Thompson suggest that scientists unwittingly succumbed to knowledge filtration that accepted current human evolutionary dogma to establish guidelines for responsible fossil interpretation. Currently accepted theory holds that "protohuman hominids evolved from apelike predecessors in Africa during the Late Pliocene and Early Pleistocene" Periods.[10] Later, modern human beings evolved from "protohuman hominids in the Late Pleistocene, in Africa, and elsewhere."[11]

Paleoanthropologists finding modern-appearing human fossil remains in Early Pleistocene or Pliocene strata, or radiometrically dating them substantially older than 100,000 years, would now consider this kind of evidence anomalous. Working from the pre-established paradigm of current theory's temporal structure, such evidence makes no sense unless dating is somehow grossly erroneous in antiquity's direction. Scientists expect fossils to fit properly on standard theory's time scale. If they do not, they must be ignored, forgotten, suppressed, or re-dated to fit their proper niche.

Over time, evolutionists subjected to knowledge filtration most earlier paleoanthropological evidence suggesting anatomically modern humans in Early Pleistocene or Pliocene strata. As a result, "most people, including professional scientists, are exposed to only a carefully edited selection of evidence supporting the currently accepted theory."[12] Contemporary scientists, finding anomalous evidence today, may in some cases risk their professional reputations if they attempt to publish their discoveries.[13] *Forbidden Archeology* documents how this filtration process has conditioned a century's understanding of human evolutionary theory.

Various techniques date fossils: radiometric methods, association with flora and fauna, morphology, nitrogen-content, fluorine content, amino acid racemization tests, and stratigraphy among others. We hear much today about radiometric testing. But the originally preferred method was based on stratigraphical geology. Stratigraphical theory argues Earth's strata were formed by sedimentary layers deposited one upon another, continuously or intermittently, the deepest being oldest.

Charles Darwin embraced Charles Lyell's uniformitarian geological model because he needed slow environmental changes for natural selection to operate. Darwinian theory's acceptance sealed the geological column's triumph. Most rocks and fossils can be dated through the superposition principle. Deeper generally means older. If this principle is rejected in total, so is evolution. This principle gave rise to the geological column, evolutionary time's basis.

Statigraphy grounded geological times largely constructed in the early nineteenth century. Quantitative assignments of ages were not initially available. One period was merely considered as prior to another. Early on, when scientists estimated Earth's age under 100,000 years, they presumed these time periods were much shorter. Radiometric dating eventually gave rise to today's immense time spans. The Cambrian Period starts 590 million years ago. More recently, the Miocene Period ranges from 25 to 5 million years ago, Pliocene from 5 to 2 million years ago, Pleistocene from 2 million to ten thousand years ago, and Holocene the last ten thousand years.[14]

Since absolute quantitative dates have varied over the last two centuries, qualitative terms describing sedimentary periods become the common coin of inter-century dating. Despite the fact that Pleistocene dates today would be far older in absolute terms than Pliocene a century ago, according to theory, genuinely Pliocene fossils will always be older than Pleistocene. In this sense, stratigraphic order is the standard for all paleoanthropological dating. Various factors may affect individual radiometric dates' accuracy. But stratigraphic order, properly analyzed, may establish actual order of relative ages. Absent signs of intrusive burial, mudslides, tectonic anomalies, or other interfering factors, Pliocene fossils should be older than Pleistocene. Deeper is usually older.

I will refer primarily to stratigraphical order to determine human chronology. Since claims that anatomically modern human beings evolved in the Late Pleistocene Period ground human evolution's current theory, we must interpret legitimate proof of such modern human beings' presence in the Early Pleistocene, Pliocene, or earlier periods as a fatal objection to present theory. Showing that evidence for earlier than Late Pleistocene presence for modern human beings is as credible as the current theory's evidence sufficiently and logically undermines the epistemological foundation of our present understanding of human evolution.

To reprise Cremo and Thompson's complex, carefully-nuanced, massive scholarly work would exceed my study's scope. Their central point, like that made by others before them, is that well-documented evidence exists suggesting that anatomically modern human beings existed with, and even before, those hominids from which current human evolution theory claims they evolved.

Cremo and Thompson address what they term "principles of epistemology" governing paleoanthropological evidence's acceptance or rejection.[15] Unlike many other experimental sciences, paleoanthropology's inferences suffer specific limitations: (1) Discoveries are rare. Scientists cannot duplicate them at will, as in conventional laboratory sciences. (2) Discoveries destroy evidence by disturbing strata and site details. (3) Once removed, fossils often tell little. The most reliable data comes from written records describing details of finds: condition of strata, evidence of intrusive burial, surrounding environment, and so forth. Sites are sometimes lost and almost always eventually destroyed. (4) Fraud sometimes occurs as Piltdown evinces. (5) Omissions of relevant data occur because beliefs, biases, or oversights often condition observations. (6) Subjective judgments may affect fossil dating methodology.

Such practical epistemological limitations place paleoanthropology in a category far removed from experimental science conceived as easily verifiable in a laboratory and universally replicable. Once human evolution's current theory became the established paradigm, paleoanthropology's inherently weaker epistemology as an historical science facilitates knowledge filtration.

This filtering process entails: (1) anomalous evidence's suppression by social processes within the relevant knowledge communities, (2) a double standard applied to evidence, and (3) morphological dating. Like Immanuel Kant's *a priori* forms of all possible cognition, these filters condition judgments that evaluate discoveries. These filters sift old and new evidence alike.

Cremo and Thompson detail how present-day paleoanthropologists apply double standards to fossil evidence. If a find conforms to standard theory, they readily accept it. But they subject anomalous evidence to scrutiny so rigorous that no find would be admitted. If they applied equal standards to standard theory fossils and anomalous fossils, (1) both would be accepted or (2) both would be rejected. If both were rejected, evolution would lose its credibility. If both were accepted, the standard theory would have to be abandoned, since this would validate true human remains before the standard theory says they began to evolve.

The practice of modern paleoanthropology has been to accept those finds consonant with current theory, while rejecting, by application of a double standard, those finds presenting true humans as pre-dating current theory's accepted time frame. Morphological dating arbitrarily assigns fossil remains to the appropriate era based on their physical appearance's expected position on standard theory's time scale, not on their strata or associated flora or fauna. Thus, we must date anatomically modern human remains to the last 100,000 years, regardless of strata or faunal or floral context.

Knowledge filtration makes an interesting thesis. But what is its proof? Extraordinary claims require extraordinary proof. Since Cremo and Thompson challenge human evolution's currently-accepted theory, their challenging evidence requires superior demonstration. Recall, though, that competent evolutionists, who knew nothing of current theory's Java-based time scale, accepted earlier than Late Pleistocene evidence of anatomically modern human remains.

My study must be brief, especially compared to *Forbidden Archeology's*. The difficulty with presenting a few outstanding cases is that they invite easy dismissal. Still, just one instance of an earlier than Late Pleistocene anatomically modern human being utterly destroys human evolution's standard scenario. I will examine the following cases to test whether epistemological weakness appears in current human evolution theory: (1) Reck's skeleton, (2) the Castenedolo skeletons, (3) the Laetoli footprints, and (4) anomalous artifacts.

Reck's skeleton, named after Berlin University anthropologist Hans Reck, was discovered in 1913. At a time when current human evolutionary theory was not the dominant paradigm, Reck discovered an anatomically modern human skeleton with a complete, although distorted, skull, located in the uppermost part of Tanzania's Olduvai Gorge's Bed II. Reck thought the strata about 800,000 years old, but modern radiometric dating places it about 1.15 million years ago.[16]

The strata so encased the skeleton that it "had to be extracted with hammers and chisels."[17] Reck initially offered careful defense against any signs of disturbance or intrusive or recent burial.[18] After re-examining the original site with Reck in 1931, famed anthropologist Louis Leakey defended the skeleton's Bed II origination, leading Leakey to conclude "that Beijing man and Java man were not direct human ancestors."[19]

The written account of the competent scientist actually unearthing the fossil usually constitutes the most reliable evidence. In his original account, Reck carefully considered and rejected the possibility of intrusive burial.[20] After careful examination of Reck's original account and a visit to the actual site in 1931, Leakey and British Museum of Natural History's A. T. Hopwood concluded that the skeleton "was native to Bed II, as originally reported by Reck" and that "the skeletal remains belonged to an anatomically modern *Homo sapiens*."[21]

In 1913, when Reck excavated the skeleton, scientists did not consider Reck's original findings anomalous. By the early 1930s the standard paradigm was well in place. Reck's skeleton became anomalous evidence, subject to various methods of suppression, including morphological dating and radical revision of evident stratigraphy. Scientists, including Reck himself, Leakey, C. Forster Cooper, D. M. S. Watson, Hopwood, and others, took opposing positions on the Bed II original burial thesis. Even Reck and Leakey finally converted to the intrusive burial thesis. Cremo and Thompson maintain Reck and Leakey "offered no satisfactory explanation for their previous opinion that the skeleton had been found in pure, unmistakable Bed II materials."[22] Scientists advanced and countered multiple arguments for various datings. This swirl of controversy manifested itself in an urgency to recalibrate the initial, most evident, carefully described and defended, stratigraphical location of Reck's skeleton. Cremo and Thompson maintain that as good, or even better, a case can be made for Reck's original judgment, a judgment made before the process of knowledge filtration took its toll.

Several measurement methods present "conflicting evidence about the age" of Castenedolo's anatomically modern human bones of two children, a man, and a woman. A 1969 radiocarbon test indicated recency in contrast to extremely ancient stratigraphy. Cremo and Thompson maintain, "The stratigraphic evidence is unusually strong in favor of a Pliocene age for the Castenedolo bones, whereas we have observed that the carbon 14 dating is far from perfect."[23]

In early 1880, geologist Giuseppe Ragazzoni excavated the Castenedolo, Italy, remains out of blue Middle Pliocene clay. In all three cases, Ragazzoni could find no signs of intrusive burial.[24] In 1883, anatomist Giuseppe Sergi examined the human remains and the site, fully confirming Ragazzoni's findings. He absolutely ruled out intrusive burial, noting that "clay from the upper surface layers, recognizable by its intense red color, would have been mixed in."[25] He maintained such strata disturbance "would not have escaped the notice of even an ordinary person what to speak of a trained geologist."[26] Contrary to the younger age of these skeletons suggested by other more doubtful dating tests, the epistemological robustness of Castenedolo's stratigraphical evidence suggests that anatomically modern humans inhabited the Middle Pliocene Period.

Even recently, knowledge filtration appears in the Laetoli footprints' case. In 1979 in Laetoli in Tanzania, Mary Leakey, wife of famed anthropologist Louis Leakey, found three individuals' footprints in volcanic ash dated at 3.6 million years old. These footprints were virtually indistinguishable from modern human footprints. Conventional anthropologists revealed their mindset when they tried to read these prints as Australopithecine, since, according to human evolution's standard theory, such were the only hominids known to exist some three million years ago. This, despite anthropologist Timothy D. White's affirmation that the prints are "like modern human footprints."[27] He calls the "external morphology" the same as a modern human being's and notes that the big toe "doesn't stick out to the side like an ape toe."[28] The Australopithecine big toe does stand out like a thumb, like the modern chimpanzee's, and quite unlike the modern human or Laetoli footprints' big toes, which are straight. The most reasonable interpretation of Pliocene anatomically modern human-looking footprints is that anatomically modern human beings caused them.

In 1956, G. H. R. von Koenigswald reported finding extremely primitive throwing balls, or bolas, in Olduvai Gorge's Bed I. He noted South American native hunters' present use of such stones, where they tie the stones together with a long leather cord and use them as a potent hurling weapon.[29]

Cremo and Thompson report that the Leakeys also found in Bed I "an apparent leather-working tool that might have been used to fashion leather cords for the bolas."[30] Early Pleistocene genuine bolas in Olduvai Gorge's Bed I argue for true human beings' intellective presence at a time when current human evolutionary theory proposes no hominid existed capable of such activity.[31] Since archeologists generally attribute some kinds of aesthetically symmetrical artifacts exclusively to anatomically modern humans, their presence in geological contexts in violation of accepted time frames for human evolution undermines current theory.

The large number of sophisticated artifacts (such as spear-heads, mortars, and pestles) and even human bones found deep inside volcanic ash-covered mine shafts and tunnels during and after the California gold rush in the last half of the nineteenth century is a major discovery.[32] State geologist J. D. Whitney reported the most significant finds.[33] The fossils appeared, in part, in auriferous gravels that lie under the Tuolumne Table Mountain. These gravels date into the Miocene Period or earlier.[34]

Smithsonian Institution's William H. Holmes's reaction to Whitney's reports manifests knowledge filtration. Holmes acknowledged that artifacts of Tertiary origin would be science's greatest marvels. Still, he maintained that, if Whitney had fully understood present human evolution theory, he would never have advanced his conclusions, despite being confronted with impressive testimony.[35] Holmes expected facts to fit current theory.

Cremo and Thompson offer many cases to show that, no matter how far back we go in paleontological time, evidence of true human beings' presence appears. This evidence deserves a hearing. A single valid piece of skeletal evidence indicating anatomically modern humans existing prior to the Late Pleistocene Period invalidates human evolution's current theory. Among *Forbidden Archeology's* extensive case documentation, I have presented evidence supporting anomalous anatomically modern human beings, Reck's skeleton, the Castenedolo remains, and the Laetoli footprints, and evidence supporting anomalous human intellective activity, such as sophisticated artifacts, appearing before the expected standard theory time-line.

If such cases were few, their enumeration might invite dismissal. *Forbidden Archeology* presents many dozens. Cremo and Thompson claim that the best proof for their thesis is a body of scientifically-reported evidence, accumulated during the last 150 years, that contradicts the current orthodox Darwinian accounts of human origins. To test that claim, nothing substitutes for careful investigation of *Forbidden Archeology's* extensive documentation.

Cremo and Thompson's deeper purpose, one which I conclude they have achieved, is to shake the epistemological foundations of current human evolutionary theory. Serious, rational doubt about human evolution's standard theory is legitimate.

Several examples, especially, display modern anthropology's double standard for acceptance and rejection of evidence: (1) the professional rejection and publication impediments that geologist Virginia Steen-McIntyre and others suffered when they attempted to report an anomalous radiometric date of 250,000 years old for highly sophisticated stone tools found at Hueyatlaco, Mexico, compared to (2) Java Man's and Peking Man's ready acceptance. Java Man's universal acceptance occurred, despite von Koenigswald's use of native collectors, Middle Pleistocene dating of fossils found lying on the surface, and inclusion of an anatomically modern human femur found 45 feet from the original cranium that is quite unlike other *Homo erectus* femurs found at Peking or Olduvai Gorge.[37] Peking Man's acceptance occurred, despite famed anthropologist Marcellin Boule's judgment that the skulls were "monkey-like" and that the real hunters, artificers, and authors of large-scale industry at the site were modern human beings.[38] The mysterious disappearance of 90 percent of all *Sinanthropus*' fossil remains during World War II has also failed to dampen support.[39]

Rejecting the equity of such double-standard science, Cremo and Thompson conclude that

> when all the available evidence is considered impartially, an evolutionary picture of human origins fails to emerge. On the one hand, if we apply the tactic of extreme skepticism equally to all available evidence, we wind up with such an insufficiency of facts that it becomes next to impossible to say anything at all about human origins. On the other hand, if we take a more liberal, yet evenhanded, approach to the totality of evidence, we are confronted with facts demonstrative of a human presence in remote geological ages, as far back as the Eocene, and even further.[40]

If anatomically modern, true human beings were present prior to current theory's time frame, what are we to make of Acheulean symmetrical toolmaking abilities attributed to Middle Pleistocene *Homo erectus*? I had predicated my conditional conclusion that Adam and Eve belonged to Middle Pleistocene *Homo erectus* upon paleoanthropologists' claims that that population fabricated artistic Acheulean handaxes, sophisticated symmetrical stone tools manifesting intellective activity. If anatomically modern human beings existed at the same time, or prior to, *Homo erectus*, my earlier inference is negated. The human species' origin becomes more mysterious.

Association of bones with tools is difficult. Skeletal remains of hominids are quite rare. Stone tools are much more common. Most sites have only stone tools, no hominid bones. The ratio of tools to bones might be tens of thousands to one. Among hominids, current theory claims only *Homo habilis*, *Homo erectus*, Neanderthals, and *Homo sapiens* are toolmakers. Standard accounts tend to assign simplest stone tools to *Homo habilis*, Acheulean symmetrical handaxes to *Homo erectus*, a more extensive toolkit to the Neanderthals, and all the way up to finely polished stones for *Homo sapiens*. The difficulty is that "anatomically modern humans are known to make and use tools of the crudest sort."[41] In principle, if modern humans coexisted with, or predated, all above-named hominids, modern humans could be the maker of all above-named tools.

Conversely, my analysis of ape-language studies and general sentient capabilities indicates that lower primates may well be responsible for simpler types of tools, short of those exhibiting aesthetic symmetry. Since the critical question focuses primarily on *Homo erectus* and Acheulean industry, the possibility that anatomically modern human beings existed in the Early Pleistocene and Pliocene Periods is sufficient to undercut the presumption that Acheulean tools belonged to *Homo erectus*. Even if a site were found with symmetrical Acheulean handaxes and a *Homo erectus* skull and no *Homo sapiens* bones locally, logic would not dictate attribution of the handaxes to *Homo erectus*. *Homo sapiens*, known from skeletal remains elsewhere, could still be the toolmaker.

Even if some *Homo erectus* skeleton is someday found, clutching a symmetrical handaxe in its bony hand, this would prove no more than your dog bringing you the evening paper in its mouth would prove that it published it! The poor *Homo erectus* might have been walking across a field, stooped to pick up a shiny human-made handaxe, and been struck by lightning. As long as it is possible that anatomically modern human beings predate, or are contemporary with, *Homo erectus*, we can attribute all intellective activity evidence to modern-appearing human beings. *Homo erectus*, then, and other proposed hominids become as relevant to human evolution as the local zoo's gorillas. They might look like us, but prove nothing more than a divine common plan or archetype. Or, as indicated earlier, relations between species might be more mysterious than supposed.

In my arguments, I generally employ the relative geological time scale, not absolute radiometric datings. I do so to avoid (1) confusion between nineteenth- and twentieth-century absolute time scales and (2) the sometimes more controversial character of radiometric dates. Earlier I dealt with ways to reconcile apparent lengthy time spans with divine revelation.

If humans did not evolve in the manner prescribed by present evolutionary theory, we must ask whether humans evolved at all. Evolutionists saw the problem. In the late nineteenth century, James Southall, speaking at the Victoria Institute in London, expressed concern that a human skull should fail to show any change from the Pliocene Period. Insufficient time would exist for human beings to develop from the lemurs of the Eocene Period. This would be fatal to human evolutionary theory.[42] If the current human evolution theory is wrong, no substitute alternative may work. One driving force to retain evolution is atheism's naturalistic presumption. Without God, life's beginnings appear unintelligible unless evolution occurred. But God's existence can be demonstrated metaphysically, independent of any perinoetic scientific disputations. Naturalism's presumption is unfounded.

Based on careful presentation of historical scientific evidence, the work of Cremo and Thompson presents a serious epistemological challenge to human evolution's current theory, perhaps to any theory of human evolution. Since human evolution has been an essential corollary to biological evolution's general theory, this raises a challenge to all current evolutionary theory. Still, the economy principle suggests some form of evolutionary process. Perhaps, maximum possible respect for creatures' secondary causality is achieved through intra-specific evolution within philosophical natural species. Again, this might indicate that relations between species are more mysteriously subordinated to an unfolding divine plan than previously supposed.

Besides theory-specific objections raised by Cremo and Thompson, many people do not realize the paucity of human evolution's evidence and controversy over its interpretation. Early hominid fossils are quite rare. The really good hominid evolution skeletal evidence up to one million years ago would fit on a few trestle tables.[43] We can lay out no exact scenario of alleged human evolution anyway because leading experts, such as the Leakeys and Johanson, often disagree among themselves. This raises doubt as to whether we can convincingly demonstrate the transitional character of any proposed intermediate form. Whether paleoanthropologists can confidently erect an entire perinoetic science of human evolutionary origins on such narrow empirical foundations remains problematic.

EPILOGUE

The origin of the human species remains mysterious. No single proposed scenario commands assent. Traditional Christian philosophy offers apodictic proof that (1) God exists, (2) He continuously conserves all finite beings in existence, (3) human beings possess spiritual and immortal intellective souls, and (4) God directly and immediately creates human souls.

Unlike theological and philosophical science, natural science's perinoetic knowledge produces no certitude while presently embracing evolutionary theory. My philosophical analysis of historical-interdisciplinary sources regarding evolutionary theory reaches these speculative conclusions: (1) Human evolution's current theory appears rationally compatible with divine revelation. (2) Still, scholarly research, such as that found in *Forbidden Archeology*, offers reasonable stratigraphic and other evidence that modern human beings predate proposed transitional hominids, such as *Homo erectus*. This presents probable cause to doubt current human evolutionary theory. (3) If current theory is doubtful, any human evolution becomes suspect, since evolutionary time required to allow an alternate scenario might be inadequate. (4) If human beings did not evolve, then: (a) Adam and Eve did not descend from transitional hominids. (b) The metaphysically impossible proposal of "intellectual emergence" becomes irrelevant. (c) *Genesis*' popular interpretation of Adam and Eve's direct divine creation, body and soul, becomes credible. (5) Should humans be of greater antiquity than supposed, no violation occurs to *Genesis*' patriarchal genealogies, since we cannot prove them to be continuous. (6) While speculation generally concedes intra-specific evolution, inter-specific evolution is suspect because of (a) evolution's historical presumption of naturalism's false premise: that no supernatural cause exists to account for living things' origins, (b) rational objections raised by Douglas Dewar, Michael J. Behe, Austin M. Woodbury, Phillip E. Johnson, and others, and (c) human evolution's doubtfulness. (7) We may never know whether human or general evolution occurred because of (a) the complexity of issues raised, (b) evolution's unscientific unfalsifiability, and (c) perinoetic knowledge's limitations.

Cosmic and human origins hypothesize four major alternatives: (1) young-Earth scientific creationism, (2) naturalism: atheistic evolution, (3) theistic evolution, and (4) progressive creationism. Sound natural science appears to refute scientific creationism. Metaphysical science refutes naturalism by showing that God exists and creates, sustains, and can directly intervene in the natural order. Theistic evolution postulates only God's direction and sustenance of an overall evolutionary process, except for direct creation of human spiritual souls (and individual miraculous events entailed in historical revelation). Evangelical theologian Bernard Ramm maintains that, while progressive creationism accepts sound modern scientific principles and conclusions, its proponents "believe in several acts of fiat creation in the history of the earth, and this clearly differentiates this view from theistic evolution."[1]

Theistic evolution and progressive creationism appear philosophically possible and scientifically defensible. Theistic evolution maximizes creatures' secondary causality by supposing that all scientifically problematic transitions, such as non-life to life or sudden appearance of major phyla, are overcome without direct divine intervention. Progressive creationism remains open to such divine intervention, postulating that life's history should reflect such discontinuities. Assuming such discontinuities must be overcome without direct divine intervention embraces naturalism's fallacy. Absent such assumptions, the *prima facie* case appears to support progressive creationism.

In the nineteenth century, militant atheists embraced Charles Darwin's *Origin of Species*, while the Church of England recoiled in horror. As evolutionary theory ascended, blind natural selection replaced William Paley's Divine Watchmaker as the explanation for cosmic natural order. Evolutionism's philosophy supplanted divine providence with pure chance and materialistic determinism. The modern mind's makers, especially Karl Marx, Charles Darwin, and Sigmund Freud, espoused an evolutionary worldview systematically excluding God.

Why would God create a world susceptible to atheistic illusions? Why would God construct the natural order so that a universal lie could replace the Divine Word? How could He expect intellective creatures to acknowledge His creative act when evolutionism's substitute religion appeared woven into the cosmic fabric? Would not divine purpose be better served by God's creative presence undeniably evinced? A universe devoid of evolution's possibility appears far better suited to beget the universal worship demanded by religion's virtue. As a Christian philosopher, I offer two responses:

First, recall that God's production of an unfolding evolutionary cosmos might be divine creative causality's most perfect expression. Evolution may be viewed as giving greater glory to the First Cause by maximizing the natural efficacy of secondary causes. God's omnipotence and omniscience might best be manifested by providentially ordaining a world so structured from its inception as to give rise through natural interaction of creatures to ever more perfect forms of reality and even of life. Appearing deceptively like naturalistic evolution, key stages in organic life's origin and development might yet require direct supernatural intervention. That some intelligent creatures should misread this natural economy as naturalistic evolution is merely a side effect of God's most perfect creative plan. A misunderstanding permitted, but not willed.

Second, God's chosen created order perfectly maximizes human freedom and responsibility. This promotes fullest development of spiritual perfection. God could have created a world so minimizing secondary causality that His presence and power would be undeniable. All creatures could have functioned as evident marionettes, virtual advertisements for constant supernatural intervention. God accomplished what Geppetto could not. The puppet Pinocchio immediately recognizes his maker. But God conceals His creative act so well that atheism is possible: the side effect of maximum secondary causality. Evolutionary materialism serves as the consensus atheistic philosophy. For seventy years, Soviet Communism promoted atheistic evolution, systematically destroying religious belief in hundreds of millions of persons worldwide.

Were God's presence as evident as the noonday sun, who would dare defy natural or revealed moral laws? Still, how meaningful would be obedience? Moral virtue's value is far greater when chosen for love of God and moral goodness, not from fear of certain punishment. Even so, hell motivates only believers, not atheists who deny its existence. As Fëdor Dostoevski aptly points out, if God does not exist, everything is permissible. Since naturalistic evolutionism is the near-universal refuge of atheism, an evolutionary world becomes a world where persons experience maximum moral freedom, including freedom to deny God's existence and moral law. Those who choose moral evil can deny the moral order's reality, just as they deny God. In seeking God's truth, the morally good person must overcome evolution's potential deception and grasp the underlying metaphysical need for a First Cause Who creates and sustains all of nature's laws, including any evolutionary mechanisms.

Since human beings are free to reject God and His moral laws by embracing evolutionary naturalism, they can also fail to attain their last end, eternal happiness united with their Creator. Why does God permit such self-demolishing liberty? God respects a spiritual creature's freedom even to allowing self-destruction. We might wish it otherwise, but God manifests His greatest glory and perfection by calling forth our greatest qualitative perfection. Free agents' greatest qualitative perfection manifests when they choose moral good while self-deceptive evil beckons. Naturalism's possibility, the unintended side effect of creatures' maximum secondary causality, offers illusory emancipation from moral constraint. A world in which evolutionary naturalism appears a speculative possibility is perfectly designed for building the greatest saints.

Speculation concerning Adam and Eve's appearance on Earth has proven challenging. Original Sin's discernment proves even more vexing. Recall, if Original Sin did not occur, Adam and Eve's wounded nature was not communicated to their descendants. No need would exist for a Redeemer. Since Jesus Christ is that Redeemer, Christianity's theological rationale depends on Original Sin's historical reality.

While paleoanthropology explores evidence for human origins through skeletal or artifact remains, can our first parent's initial rebellion against God be empirically discerned? The correlative rebellion of human passion against reason's rule in present human beings manifests Original Sin's effects, but does not directly evince the sin itself. The essence of sin, as a perverted will act, leaves no fossil record trace. Does this fact preclude all empirical evidence of Original Sin? Perhaps not.

Really good early hominid skeletal evidence is so rare that it would fit on a few trestle tables. A single skeleton, even small parts of one, can determine the course of paleoanthropological speculation. In Java, a cranium fragment, a tooth, and an anatomically modern femur gave rise to current human evolutionary theory. If correctly dated as late Early Pleistocene, Reck's anatomically modern human skeleton could wreck that same theory. So could a single Middle Pliocene Castenedolo anatomically modern human skeleton. Interpretative analysis of single primate body remnants can produce or destroy whole theories concerning human origins.

Natural scientists apply physical laws to fossils producing culture-shaking interpretations of their meaning and significance. Universal physical law which guarantees constancy of radioactive decay rates is essential to radiometric dating accuracy. Atheistic naturalism allows no exception to physical laws because no supernatural power exists to suspend them. The most evident universal physical law concerning human remains is that death inevitably leads to bodily corruption. Yet, though little known, the bodies of at least one-hundred deceased Catholic Christians, who are in various canonization process stages, violate this universal physical law.[2] Specimens of bodies that have been deliberately or accidentally preserved "without exception were found stiff, discolored, and skeletal."[3] But the "greater majority" of these Christian incorruptibles "were never embalmed or treated in any manner, yet most were found lifelike, flexible, and sweetly scented many years after death."[4] If we can found perinoetic theories of human origins upon fragments of skeletal remains enigmatically deposited in ancient sediments, equity requires rational examination of such clearly preternatural phenomena as bodily remains that violate natural physical laws. I focus, especially, upon the body of Bernadette Soubirous, who, in life, profoundly encountered the theological doctrine of Original Sin.

Worldwide fame attends the Catholic shrine at Lourdes, France, known for the more than two thousand medically-inexplicable physical cures certified by the self-directing, self-supporting Medical Bureau. Of these cures, a mere 64 have been approved by the Church as of 1979 because of more exacting theological scrutiny. Far less known is the Convent of Saint Gildard at Nevers, France, where Bernadette Soubirous spent her last thirteen years of Earthly life in virtual anonymity. On 16 April 1879, at age 35, she died there from tuberculosis of the knee. Her body was buried, without embalming or any other artificial means of preservation, in St. Joseph's Chapel inside the convent. In 1909, during her canonization process, her body was exhumed and found to be perfectly incorrupt. Her lips and fingertips show the fullness of life and her general appearance has been described as "beautiful." On 8 December 1933, the Catholic Church proclaimed Bernadette Soubirous a saint. Her bodily remains now reside in a gold and glass coffin exposed to public view in a side chapel at the convent motherhouse at Nevers. Thousands have personally seen her incorrupt body. Its image has been telecast throughout the world. Many books contain its photograph.[5] My concern is the theological significance of this evidently preternatural phenomenon.

Unless anti-religious knowledge filtration supervenes, such preternatural preservation constitutes data present in the physical world whose intelligibility we should consider as seriously as any paleoanthropological fossil discovery. Whether this phenomenon is supernatural, not merely preternatural, but caused by God alone, I leave to appropriate theological inference. For those who regard this incorruption as supernatural in nature and origin, I suggest the following: Bernadette was an ignorant, peasant visionary who affirmed that the lady dressed in white whom she saw at the Lourdes Grotto proclaimed, "I am the Immaculate Conception." Only four years earlier, on 8 December 1854, Pope Pius IX solemnly proclaimed the dogma of the Immaculate Conception. The Blessed Virgin's Immaculate Conception means that from her conception's first moment in her mother's womb Mary was preserved free from Original Sin. If Original Sin is not historical reality, that dogma and Bernadette's subsequent statement about the vision become unintelligible. Conversely, if her body's preservation, suspending natural physical laws is viewed as a sign of God's approval of her life and works, then Bernadette's affirmation of the Immaculate Conception is simultaneously an affirmation of the reality of Original Sin and of the individual persons who committed that sin, Adam and Eve. Bernadette's body manifests lasting visible evidence of authentic retrospective prophecy concerning Original Sin.

As we enter the Third Millennium, the empirically verifiable bodily remains of St. Bernadette Soubirous lead us back to our distant past, affirming a mysterious, but specific, time and place of human origins for our literal, Biblical first parents, Adam and Eve.

Today's universal assumption of evolution's truth makes people comfortable with the notion that Adam and Eve's story is pure mythology. Perhaps, that is why G. K. Chesterton had to remind us that Christianity is a myth that is true.[6]

NOTES

Preface

1. Raymond J. Nogar, *The Wisdom of Evolution* (Garden City, N.Y.: Doubleday & Co., Inc., 1963), p. 17.

2. Sir Arthur Eddington, *The Nature of the Physical World* (New York: The Macmillan Company and Cambridge, England: The University Press, 1929), p. 259.

Chapter One

1. Aristotle, *Metaphysics* Bk. 1, 5, 985b24, in *The Basic Works of Aristotle*, ed. Richard McKeon (New York: Random House, 1941), p. 698.

2. *Ibid.*, 987a19, p. 700.

3. *Ibid.*, 987a20-22.

4. *Ibid.*, 6, 987b0-4, pp. 700-701.

5. *Ibid.*, 987b6-7, p. 701.

6. *Ibid.*, 987b8-13.

7. Hippolytus, *Hippolyti Philosophumena* 6, Dox. 559 as cited in Milton C. Nahm, *Selections from Early Greek Philosophy* (New York: Appleton-Century-Crofts, 1964), p. 42.

8. R. F. Baum, "Coming to Grips with Darwin," in *The Intercollegiate Review*, 11:1 (Fall, 1975), p. 19.

9. Immanuel Velikovsky, *Earth in Upheaval* (New York: Dell Publishing Co., Inc., 1955), p. 235.

10. George Gaylord Simpson, "Organic Evolution," in *Collier's Encyclopedia*, Vol. 9, ed. Louis Shores (Canada: P.F. Collier & Son, Ltd., 1973), p. 481.

11. Roy A. Gallant, "To Hell with Evolution," in *Science and Creationism*, ed. Ashley Montagu (Oxford, New York, Toronto, Melbourne: Oxford University Press, 1984), p. 285.

12. Laurie R. Godfrey, "Scientific Creationism: The Art of Distortion," in *Science and Creationism*, p. 172.

13. *Ibid.*, p. 260.

148   *ORIGIN OF THE HUMAN SPECIES*

14. Roger J. Cuffey, "Paleontologic Evidence and Organic Evolution," in *Science and Creationism*, p. 257.

15. *Ibid.*, p. 268.

16. *Ibid.*, p. 258.

17. A. N. Field, *The Evolution Hoax Exposed* (Rockford, Ill.: Tan Books and Publishers, 1971), p. 27.

18. *Ibid.*

19. Douglas Dewar, *More Difficulties of the Evolution Theory*, p. 144, as cited by Field, *The Evolution Hoax Exposed*, p. 27.

20. *Ibid.*

21. Stephen Jay Gould, *Hen's Teeth and Horse's Toes* (New York-London: W. W. Norton & Co., 1983), p. 260.

22. Charles Darwin, *On the Origin of Species* (Cambridge, Mass. and London, England: Harvard University Press, 1964), p. 302.

23. T. A. Goudge, *The Ascent of Life* (Toronto: University of Toronto Press, 1961), p. 36.

24. Field, *The Evolution Hoax Exposed*, p. 30.

25. Dewar, *Challenge to Evolutionists*, pp. 52-57, as cited by Field, *The Evolution Hoax Exposed*, p. 31.

26. Michael J. Behe, *Darwin's Black Box* (New York: The Free Press), 1996.

27. *Ibid.*, p. 40.

28. *Ibid.*

29. Joel L. Cracraft, "Systematics, Comparative Biology, and Creationism," in *Scientists Confront Creationism*, ed. Laurie R. Godfrey (New York-London: W. W. Norton & Company, 1983), p. 189.

30. Raymond J. Nogar, *The Wisdom of Evolution* (Garden City, N.Y.: Doubleday & Company, Inc., 1963), p. 64.

31. *Ibid.*, p. 30.

32. *Ibid.*, p. 31.

33. *Ibid.*

34. *Evolution after Darwin*, ed. Sol Tax (Chicago: University of Chicago Press, 1960).

35. Nogar, *The Wisdom of Evolution*, p. 32.

36. *Ibid.*, p. 39.

37. *Ibid.*, pp. 70-71.

38. *Ibid.*, p. 123.

39. *Ibid.*, pp. 121-123.

40. *Ibid.*, pp. 124-125.

41. *Ibid.*, p. 38.

42. *Ibid.*

43. *Ibid.*, p. 143.

44. *Ibid.*, pp. 92-93.

45. Sidney W. Fox, "Creationism and Evolutionary Protobiogenesis," in *Science and Creationism*, p. 201.

46. Nogar, *The Wisdom of Evolution,* p. 143.

47. *Ibid.*, p. 39.

48. Austin M. Woodbury, *Philosophical Psychology* (Sydney, N.S.W.: Aquinas Academy, unpublished manuscript, 1945), p. 61.

49. *Ibid.*

50. *Ibid.*

51. *Ibid.*

52. *Ibid.*

53. Nogar, *The Wisdom of Evolution*, p. 66.

54. *Ibid.*, p. 69.

55. Woodbury, *Philosophical Psychology*, p. 61.

56. Nogar, *The Wisdom of Evolution*, p. 45.

57. St. Thomas Aquinas, *Summa theologiae,* 1, q. 22, a. 3, c (Ottawa: Collège Dominicain d'Ottawa, 1941-1945).

58. Nogar, *The Wisdom of Evolution*, pp. 69-70.

59. *Ibid.*, p. 70.

60. Etienne Gilson, *The Spirit of Mediaeval Philosophy*, trans. A. H. C. Downes (New York: Charles Scribner's Sons, 1940), pp. 144-145.

61. Phillip E. Johnson, *Darwin on Trial* (Washington: Regnery Gateway, 1991).

62. *Ibid.*, p. 125.

63. *Ibid.*

64. *Ibid.*, pp. 114-115.

65. *Ibid.*, p. 50.

66. *Ibid.*, p. 51.

67. *Ibid.*, pp. 75-79.

68. *Ibid.*, p. 52.

69. *Ibid.*

70. *Ibid.*, pp. 112, 120, and 153-154.

71. *Ibid.*, p. 151.

72. *Ibid.*, p. 91.

73. *Ibid.*, p. 67.

74. *Ibid.*, p. 64.

75. *Ibid.*, p. 139.

76. *Ibid.*, p. 89.

77. Michael Denton, *Evolution: A Theory in Crisis* (Bethesda, Md.: Adler & Adler, Publishers, Inc., 1985), p. 99.

78. *Ibid.*, pp. 100-117.

79. *Ibid.*, p. 193.

80. *Ibid.*, p. 200.

81. *Ibid.*, pp. 194-195, 201.

Chapter Two

1. St. Thomas Aquinas, *Summa theologiae*, 1, q. 2, a. 2, ad 1 (Ottawa: Collège Dominicain d'Ottawa, 1941-45).

2. John N. Deely, *The Philosophical Dimensions of the Origin of Species* (Chicago: Institute for Philosophical Research, 1969), p. 77.

3. Ernst Mayr, *Animal Species and Evolution* (Cambridge, Mass. and London, England: The Belknap Press of Harvard University Press, 1963), p. 5.

4. *Ibid.*, p. 31.

5. *Ibid.*, pp. 31-34 ff.

6. *Ibid.*, p. 20.

7. *Ibid.*, p. 28.

8. *Ibid.*, p. 27.

9. *Ibid.*, p. 23.

10. Jacques Maritain, *The Degrees of Knowledge*, trans. Gerald B. Phelan (New York: Charles Scribner's Sons, 1959), pp. 202-210.

11. Mayr, *Animal Species and Evolution*, p. 23.

12. *Ibid.*, pp. 16-17.

13. Mayr, *The Species Problem* (Washington: American Association for the Advancement of Science, 1957), p. 17.

14. Raymond J. Nogar, *The Wisdom of Evolution* (Garden City, N.Y.: Doubleday & Company, Inc., 1963), p. 326.

15. *Ibid.*, p. 330.

16. *Ibid.*, p. 332.

17. *Ibid.*, p. 331.

18. *Ibid.*, p. 324.

19. *Ibid.*, pp. 324-335.

20. *Ibid.*, p. 324.

21. *Ibid.*, p. 332.

22. *Ibid.*, p. 328.

23. *Ibid.*, p. 332.

24. Austin M. Woodbury, *Philosophical Psychology* (Sydney, N.S.W.: Aquinas Academy, unpublished manuscript, 1945), p 50.

25. *Ibid.*, p. 49.

26. *Ibid.*, p. 50.

27. *Ibid.*, p. 63.

28. *Ibid.*, p. 58.

29. Nogar, *The Wisdom of Evolution*, p.125.

30. Roger J. Cuffey, "Paleontologic Evidence and Organic Evolution," in *Science and Creationism*, ed. Ashley Montagu (Oxford, New York, Toronto, Melbourne: Oxford University Press, 1984), p. 257.

31. *Ibid.*, p. 262.

32. Woodbury, *Philosophical Psychology*, p. 57.

33. Etienne Gilson, *The Philosophy of St. Thomas Aquinas (Le Thomisme)*, trans. Edward Bullough (Cambridge: W. Heffer & Sons, Ltd., 1929), p. 154; see, also, Aquinas, *Summa contra gentiles*, Bk. 4, Ch. 11, Vols. 13-15 (ed. Leonine: Rome, 1918).

34. Aquinas, *Summa theologiae,* 1, q. 76, a. 3, c (Ottawa: Collège Dominicain d'Ottawa, 1941-1945).

35. *Ibid.*

36. Woodbury, *Philosophical Psychology*, p. 58.

37. *Ibid.*

38. *Ibid.*, p. 62.

39. *Ibid.*, p. 57.

40. *Ibid.*, pp. 57-58.

41. *Ibid.*

42. *Ibid.*, p. 63.

43. *Ibid.*

Chapter Three

1. Austin M. Woodbury, *Philosophical Psychology* (Sydney, N.S.W.: Aquinas Academy, unpublished manuscript, 1945), p. 57.

2. *Ibid.*, pp. 59-62.

3. *Ibid.*, p. 60.

4. *Ibid.*

5. *Ibid.*, p. 63.

6. Roger J. Cuffey, "Paleontologic Evidence and Organic Evolution," in *Science and Creationism*, ed. Ashley Montagu (Oxford, New York, Toronto, Melbourne: Oxford University Press, 1984), pp. 255-281; C. Loring Brace, "Humans in Time and Space," in *Scientists Confront Creationism*, ed. Laurie R. Godfrey (New York and London: W. W. Norton & Company, 1983), pp. 245-282.

7. Woodbury, *Philosophical Psychology*, p. 59.

8. *Ibid.*

9. St. Thomas Aquinas, *Summa theologiae*, 1, q. 110, a. 4, ad 2 (Ottawa: Collège Dominicain d'Ottawa, 1941-1945); see, also, *ibid.*, q. 105, aa. 7, 8.

10. Woodbury, *Philosophical Psychology*, p. 59.

11. Aquinas, *Summa theologiae*, 1, q. 110, a. 4.

12. Dennis Bonnette, "How 'Creation' Implies God," in *Faith & Reason*, 11:3-4 (1985), pp. 250-263.

13. Woodbury, *Philosophical Psychology*, p. 57.

14. *Ibid.*

15. Aquinas, *Summa contra gentiles*, Bk. 3, Ch. 1-5, Vols. 13-15 (ed. Leonine: Rome, 1918); Jacques Maritain, *A Preface to Metaphysics* (New York: Sheed & Ward, 1939), pp. 105-131; Reginald Garrigou-Lagrange, *God: His Existence and His Nature*, trans. Dom Bede Rose (St. Louis-London: B. Herder Book Co., 1934), pp. 199-205.

16. Woodbury, *Philosophical Psychology*, p. 59.

17. *Ibid.*

18. *Ibid.*

19. *Ibid.*, pp. 60-61.

20. *Ibid.*, p. 61.

21. *Ibid.*, p. 59.

22. Raymond J. Nogar, *The Wisdom of Evolution* (Garden City, N.Y.: Doubleday & Company, Inc., 1963), p. 70.

23. *Ibid.*, pp. 69-70.

24. Bonnette, "Hylomorphism, Positivism, and the Question of When Human Life Begins," in *Social Justice Review*, 72:3-4 (March-April 1981), pp. 42-45.

25. Nogar, *The Wisdom of Evolution*, pp. 224-225.

26. T. A. Goudge, *The Ascent of Life* (Toronto: University of Toronto Press, 1961), p. 37.

27. *Ibid.*

28. Robert Jastrow, *God and the Astronomers* (New York: W. W. Norton and Co., Inc., 1978), pp.11-12.

29. Goudge, *The Ascent of Life*, p. 37.

30. *Ibid.*

31. Woodbury, *Philosophical Psychology*, p. 46.

32. *Ibid.*, p. 47.

33. *Ibid.*, p. 59.

34. *Ibid.*

35. *Ibid.*

36. John W. Patterson, "Thermodynamics and Evolution," in *Scientists Confront Creationism*, p. 110.

37. Austin M. Woodbury, *Cosmology* (Sydney, N.S.W.: Aquinas Academy, unpublished manuscript, 1949), p. 67.

38. *Ibid.*, p. 67.

39. *Ibid.*, p. 68.

40. Patterson, "Thermodynamics and Evolution," in *Scientists Confront Creationism*, p. 104.

41. *Ibid.*, p. 105.

42. John N. Deely, *The Philosophical Dimensions of the Origin of Species* (Chicago: Institute for Philosophical Research, 1969), p. 324.

43. *Ibid.*, pp. 324-325; See, also, pp. 317-331.

44. *Ibid.*, p. 324.

45. *Ibid.*, p. 321.

46. Jacques Maritain, *Approaches to God* (New York: Collier Books, 1962), p. 57.

47. Jacques Maritain, *A Preface to Metaphysics* (New York: Sheed & Ward, 1939), pp. 141-151.

48. Woodbury, *Cosmology*, p. 68.

49. *Ibid.*, p. 67.

50. Aquinas, *Summa theologiae*, 1, q. 115, a. 6, c.

51. *Ibid.*

52. Maritain, *A Preface to Metaphysics*, p. 148

53. *Ibid.*, pp. 141-151. See, also, Maritain, *Approaches to God*, pp. 56-62; Bonnette, *Aquinas' Proofs for God's Existence* (The Hague: Martinus-Nijhoff, 1972), pp. 157-175; Reginald Garrigou-Lagrange, *God: His Existence and His Nature*, trans. Dom Bede Rose (St. Louis: B. Herder Book Co., 1939), Vol. 1, pp. 345-372.

54. *Romans* 1, 20.

55. Bonnette, "How 'Creation' Implies God," in *Faith & Reason*, 11:3-4 (1985), pp. 250-263 and "A Variation on the First Way of St. Thomas Aquinas," in *Faith & Reason*, 8:1 (Spring 1982), pp. 34-56.

56. Sidney W. Fox, "Creationism and Evolutionary Protobiogenesis," in *Science and Creationism*, ed. Ashley Montagu (Oxford, New York, Toronto, Melbourne: Oxford University Press, 1984), p. 210.

57. *Ibid.*, pp. 194-231, esp. p. 211.

58. *Ibid.*, p. 227.

59. *Ibid.*, p. 228.

60. *Ibid.*, p. 215.

61. *Ibid.*, p. 229.

62. *Ibid.*, p. 230.

63. *Ibid.*

64. Maritain, *Approaches to God*, pp. 59-60.

65. Aquinas, *Summa theologiae*, 1, q. 2, a. 3. c and *Summa contra gentiles*, Bk. 1, Ch. 13, Vols. 13-15 (ed. Leonine: Rome, 1918).

66. Sidney W. Fox, "Creationism and Evolutionary Protobiogenesis," in *Science and Creationism*, p. 204.

67. R. F. Baum, "Coming to Grips with Darwin," in *The Intercollegiate Review*, 11:1 (Fall, 1975), p. 18.

68. Fox, "Creationism and Evolutionary Protobiogenesis," in *Science and Creationism*, p. 229.

69. *Ibid.*

70. Woodbury, *Philosophical Psychology*, p. 59.

71. *Ibid.*

72. *Ibid.*, p. 60.

73. *Ibid.*

74. *Ibid.*, p. 49.

75. *Ibid.*, p. 53.

76. Brother Benignus Gerrity, *Nature, Knowledge, and God* (Milwaukee: The Bruce Publishing Company, 1947), pp. 501-502.

77. *Ibid.*, p. 501.

78. Aquinas, *Summa contra gentiles*, Bk. 3, Ch. 23, Vols. 13-15.

79. *Ibid.*

80. Gerrity, *Nature, Knowledge, and God*, p. 501.

81. *Ibid.*

82. *Ibid.*

83. *Ibid.*, p 502.

84. *Ibid.*

Chapter Four

1. Austin M. Woodbury, *Philosophical Psychology* (Sydney, N.S.W.: Aquinas Academy, unpublished manuscript, 1951), Bk. 3, Ch. 1, pp. 383-388.

2. *Ibid.*, p. 382.

3. *Ibid.*, p. 383.

4. *Ibid.*, pp. 383-384.

5. *Ibid.*, p. 384.

6. *Ibid.*

7. Rémy Collin, "*Réflexions sur le psychisme*," in *Cahiers de philosophie de la nature*, Bk. 3, p. 118, as cited by Woodbury, *Philosophical Psychology*, 1951, Bk. 3, Ch. 1, pp. 384-385.

8. Woodbury, *Philosophical Psychology*, 1951, Bk. 3, Ch. 1, p. 385.

9. *Ibid.*, p. 386.

10. *Ibid.*, p. 387.

11. Austin M. Woodbury, *Philosophical Psychology* (Sydney, N.S.W.: Aquinas Academy, unpublished manuscript, 1945), p. 59.

12. Austin M. Woodbury, *Cosmology* (Sydney, N.S.W.: Aquinas Academy, unpublished manuscript, 1949), p. 67.

13. Heini K. P. Hediger, "The Clever Hans Phenomenon from an Animal Psychologist's Point of View," in *The Clever Hans Phenomenon: Communication with Horses, Whales, Apes, and People*, ed. Thomas A. Sebeok and Robert Rosenthal (New York: New York Academy of Sciences, 1981), p. 12.

Chapter Five

1. Herbert S. Terrace, "How Nim Chimpsky Changed My Mind," in *Psychology Today*, 13:6 (November 1979), p. 67.

2. Suzanne Chevalier-Skolnikoff, "The Clever Hans Phenomenon, Cuing, and Ape Signing: A Piagetian Analysis of Methods for Instructing Animals," in *The Clever Hans Phenomenon: Communication with Horses, Whales, Apes, and People*, ed. Thomas A. Sebeok and Robert Rosenthal (New York: New York Academy of Sciences, 1981), pp. 89-90.

3. *Ibid.*, p. 83.

4. Thomas A. Sebeok and Jean Umiker-Sebeok, "Performing Animals: Secrets of the Trade," in *Psychology Today*, 13:6 (November 1979), p. 91.

5. *Ibid.*, p. 81.

6. Stephen Walker, *Animal Thought* (London, Boston, Melbourne, Henley: Routledge & Kegan Paul, Ltd., 1983), pp. 373-374.

7. Duane M. Rumbaugh, "Who Feeds Clever Hans?", in *The Clever Hans Phenomenon*, pp. 26-34. See, also, E. Sue Savage-Rumbaugh, "Can Apes Use Symbols to Represent Their World?", in *The Clever Hans Phenomenon*, pp. 35-59.

8. Sebeok and Umiker-Sebeok, "Performing Animals: Secrets of the Trade," in *Psychology Today*, 13:6 (November 1979), p. 91.

9. Walker, *Animal Thought*, p. 373.

10. *Ibid.*, pp. 365-367.

11. *Ibid.*, p. 369.

12. *Ibid.*, p. 370.

13. *Ibid.*, pp. 370-371.

14. *Ibid.*, p. 377.

15. *Ibid.*, pp. 377-378.

16. *Ibid.*, p. 378.

17. *Ibid.*, p. 379.

18. *Ibid.*, p. 387.

19. *Ibid.*

20. *Ibid.*, p. 357.

21. *Ibid.*

22. Heini K. P. Hediger, "The Clever Hans Phenomenon from an Animal Psychologist's Point of View," in *The Clever Hans Phenomenon*, p. 5.

23. *Ibid.*

24. *Ibid.*

25. Aristotle, *On Interpretation*, 1, 16a3-8, in *The Basic Works of Aristotle*, ed. Richard McKeon (New York: Random House, 1941), p. 40.

26. Hediger, "The Clever Hans Phenomenon from an Animal Psychologist's Point of View," p. 9.

27. *Ibid.*

28. *Ibid.*, p. 9.

29. *Ibid.*

30. *Ibid.*, p. 13.

31. *Ibid.*

32. *Ibid.*, p. 14.

33. *Ibid.*, p. 16.

34. Sebeok and Umiker-Sebeok, "Performing Animals: Secrets of the Trade," in *Psychology Today*, p. 91.

35. Walker, *Animal Thought*, pp. 352-381.

36. Hediger, "The Clever Hans Phenomenon from an Animal Psychologist's Point of View," p. 14.

37. Aristotle, *History of Animals*, Bk. 8, 1, 589a3-589a9, in *The Basic Works of Aristotle*, ed. Richard McKeon (New York: Random House, 1941), p. 636.

38. Chevalier-Skolnikoff, "The Clever Hans Phenomenon, Cuing, and Ape Signing: A Piagetian Analysis of Methods for Instructing Animals," in *The Clever Hans Phenomenon*, p. 83.

39. *Ibid.*

40. *Ibid.*, p. 84.

41. Walker, *Animal Thought*, p. 374.

42. Chevalier-Skolnikoff, "The Clever Hans Phenomenon, Cuing, and Ape Signing," pp. 89-90.

43. *Ibid.*, p. 86.

44. Terrace, "A Report to an Academy, 1980," in *The Clever Hans Phenomenon*, p. 95.

45. *Ibid.*

46. *Ibid.*, p. 96.

47. *Ibid.*, p. 97.

48. *Ibid.*, p. 103.

49. *Ibid.*, pp. 107-108.

50. *Ibid.*, p. 108.

51. Mark S. Seidenberg and Laura A. Petitto, "Ape Signing: Problems of Method and Interpretation," in *The Clever Hans Phenomenon*, p. 116.

52. Terrace, "A Report to an Academy, 1980," p. 109.

53. *Ibid.*, pp. 109-110.

54. Seidenberg and Petitto, "Ape Signing: Problems of Method and Interpretation," in *The Clever Hans Phenomenon*, p. 116.

55. *Ibid.*, pp. 121-122.

56. *Ibid.*, p. 123.

57. *Ibid.*, p. 127.

58. Terrace, "In the Beginning Was the 'Name,'" in *American Psychologist*, 40:9 (September 1985), p. 1011.

59. *Ibid.*, p. 1017.

60. *Ibid.*, p. 1022.

61. *Ibid.*, p. 1024.

62. Walker, *Animal Thought*, p. 9.

63. *Ibid.*, pp. 364-370, 373.

64. *Ibid.*, pp. 369-370.

65. *Ibid.*, pp. 370-371.

66. *Ibid.*, pp. 10-11.

67. Stanley L. Jaki, *Brain, Mind, and Computers* (New York: Herder and Herder, Inc., 1969), p. 110.

68. *Ibid.*, p. 115.

69. Hediger, "The Clever Hans Phenomenon from an Animal Psychologist's Point of View," p. 5.

70. St. Thomas Aquinas, *Quaestiones disputatae de potentia dei*, q. 10, a. 4, c, in St. Thomas Aquinas, *Quaestiones disputatae*, Vol. 22, eds. P. Bazzi *et al.*, 10th edition (Turin-Rome: Marietti, 1965).

71. Aquinas, *Summa contra gentiles*, Bk. 3, Ch. 23, Vols. 13-15 (ed. Leonine: Rome, 1918).

72. Aristotle, *Physics*, Bk. 2, 1, 192b22-23, in *The Basic Works of Aristotle*, ed. Richard McKeon (New York: Random House, 1941), p. 236.

73. Jaki, *Brain, Mind, and Computers*, p. 214.

74. *Ibid.*, p. 215.

75. *Ibid.*, p. 216.

76. *Ibid.*, pp. 220-221.

77. *Ibid.*

78. Paul Bouissac, "Behavior in Context: In What Sense Is a Circus Animal Performing?", in *The Clever Hans Phenomenon*, p. 24.

79. Austin M. Woodbury, *Natural Philosophy*, Treatise Three, *Psychology*, Bk. 3, Ch. 40, Art. 7 (Sydney: Aquinas Academy, unpublished manuscript, 1951) pp. 432-465.

80. *Ibid.*, p. 437.

81. *Ibid.*, p. 438.

82. *Ibid.*

83. *Ibid.*, p. 441.

84. *Ibid.*, p. 443.

85. *Ibid.*, p. 444.

86. *Ibid.*

87. *Ibid.*, p. 445.

88. *Ibid.*, p. 447.

89. *Ibid.*, p. 448.

90. *Ibid.*

91. *Ibid.*, p. 433.

92. David Hume, *A Treatise of Human Nature*, Vol. 1, Bk. 1, Sect. 1 (London: J.M. Dent & Sons, Ltd., 1956), pp. 11-16.

93. Woodbury, *Natural Philosophy*, p. 434.

Chapter Six

1. St. Thomas Aquinas, *Summa theologiae*, 1, q. 75, a. 2, c (Ottawa: Collège Dominicain d'Ottawa, 1941-1945).

2. Austin M. Woodbury, *Philosophical Psychology* (Sydney, N.S.W.: Aquinas Academy, unpublished manuscript, 1945), pp. 830-839.

3. *Ibid.*, pp. 827-830.

4. *Ibid.*, pp. 840-845.

5. Aquinas, *Quaestiones disputatae de veritate*, q. 2, a. 2, Vol. 1, ed. R. Spiazzi (Turin-Rome: Marietti, 1964).

6. Woodbury, *Philosophical Psychology*, p. 840.

7. *Ibid.*

8. Aquinas, *Summa theologiae*, 1, q. 90, a. 2, ad 2.

9. *Ibid.*, 1, q. 75, a. 2, c.

10. *Ibid.*, 1, q. 90, a. 2, c.

11. *Ibid.*, 1, q. 90, a. 3, c.

12. Aquinas, *Summa contra gentiles*, Bk. 2, Ch. 87, Vols. 13-15 (ed. Leonine: Rome, 1918).

13. *Ibid.*, p. 21.

14. Aquinas, *Summa theologiae*, 1, q. 45, a. 5, ad 3.

15. Dennis Bonnette, "How 'Creation' Implies God," in *Faith & Reason*, 11:3-4 (1985), pp. 250-263.

Chapter Seven

1. Austin M. Woodbury, *Philosophical Psychology* (Sydney, N.S.W.: Aquinas Academy, unpublished manuscript, 1945), p. 59.

2. John E. Pfeiffer, *The Emergence of Man* (New York, Evanston, and London: Harper & Row, Publishers, 1969), p. 437.

3. *Ibid.*

4. Sir Francis Graham-Smith and Sir Bernard Lovell, *Pathways to the Universe* (Cambridge, England: Cambridge University Press, 1988), pp. 218-225.

5. *Ibid.*, p. 222.

6. *Ibid.*, p. 221.

7. *Ibid.*, p. 218.

8. *Ibid.*, p. 220.

9. Sir Bernard Lovell, *In the Center of Immensities* (New York, Hagerstown, San Francisco, London: Harper & Row, Publishers, 1978), p. 63.

10. *Ibid.*

11. Graham-Smith and Lovell, *Pathways to the Universe*, p. 220.

12. Lovell, *In the Center of Immensities*, p. 64.

13. Graham-Smith and Lovell, *Pathways to the Universe*, p. 220.

14. *Ibid.*, pp. 220-221.

15. *Ibid.*

16. *Ibid.*, p. 219.

17. *Ibid.*

18. Lovell, *In the Center of Immensities*, pp. 122-123.

19. *Ibid.*

20. *Ibid.*, p. 122.

21. Graham-Smith and Lovell, *Pathways to the Universe*, p. 217.

22. Jacques Maritain, *A Preface to Metaphysics* (New York: Sheed & Ward, 1939), p. 148.

23. Lecomte du Noüy, *The Road to Reason* (New York and Toronto: Longmans, Green and Co., 1948), pp. 124-125.

24. Graham-Smith and Lovell, *Pathways to the Universe*, p. 221.

25. *Ibid.*, p. 219.

26. Sidney W. Fox, "Creationism and Evolutionary Protobiogenesis," in *Science and Creationism*, ed. Ashley Montagu (Oxford, New York, Toronto, Melbourne: Oxford University Press, 1984), p. 200.

27. *Ibid.*, pp. 202-203.

28. *Ibid.*, p. 215.

29. Graham-Smith and Lovell, *Pathways to the Universe*, pp. 220-221.

30. Sidney W. Fox, "Creationism and Evolutionary Protobiogenesis," in *Science and Creationism*, p. 218.

31. *Ibid.*, p. 227.

32. John W. Patterson, "Thermodynamics and Evolution," in *Scientists Confront Creationism*, ed. Laurie R. Godfrey (New York and London: W. W. Norton & Company, 1983), p. 111.

33. *Ibid.*

34. J. W. G. Johnson, *The Crumbling Theory of Evolution* (Los Angeles: Perpetual Eucharistic Adoration, Inc., 1986), pp. 94-95.

35. *Ibid.*

36. Michael J. Behe, *Darwin's Black Box* (New York: The Free Press, 1996), p. 170.

37. *Ibid.*, pp. 166-167.

38. *Ibid.*, p. 171.

39. *Ibid.*, pp. 172-173.

40. Walter T. Brown, Jr., *In the Beginning* (Phoenix, Ariz.: Center for Scientific Creation, 1989), p. 7.

41. *Ibid.*, p. 41.

42. *Ibid.*, p. 8.

43. *Ibid.*

44. Russell F. Doolittle, "Probability and the Origin of Life," in *Scientists Confront Creationism*, p. 89.

45. Brown, Jr., *In the Beginning*, p. 8.

46. *Ibid*.

47. A. E. Ringwood, *Origin of the Earth and Moon* (New York, Heidelberg, Berlin: Springer-Verlag, 1979), p. 71.

48. T. Encrenaz and J.-P. Bibring, *The Solar System* (Berlin, Heidelberg, New York, London-Paris, Tokyo, Hong Kong: Springer-Verlag, 1990), p. 311.

49. Hugh Ross, *The Creator and the Cosmos* (Colorado Springs, Col.: Nav-Press, 1995), pp. 154-155.

50. Encrenaz and Bibring, *The Solar System*, pp. 311-312.

51. *Ibid*.

Chapter Eight

1. St. Thomas Aquinas, *Summa theologiae*, 1, q. 75, a. 5, ad 1 (Ottawa: Collège Dominicain d'Ottawa, 1941-1945).

2. *Ibid*., q. 76, a. 5, ob. 1.

3. Reginald Garrigou-Lagrange, *God: His Existence and His Nature* (St. Louis and London: B. Herder Book Co., 1936), Vol. 2, p. 554.

4. Charles B. Crowley, *Aristotelian-Thomistic Philosophy of Measure and the International System of Units (SI)*, ed. Peter A. Redpath (Lanham, New York, London: University Press of America, 1996), pp. 249-261.

5. Austin M. Woodbury, *Philosophical Psychology* (Sydney, N.S.W.: Aquinas Academy, unpublished manuscript, 1945), p. 60.

6. John N. Deely, *The Philosophical Dimensions of the Origin of Species* (Institute for Philosophical Research: Chicago, 1969), p. 306

7. St. Thomas Aquinas, *Summa theologiae*, 1-2, q. 52, a. 1, c.

8. *Ibid*., 1, q. 76, a. 4, ad 4.

9. *Ibid*., 1-2, q. 52, a. l, c.

10. *Ibid*.

11. *Ibid*.

Chapter Nine

1. Henricus Denzinger, *Enchiridion symbolorum*, 29th edition, 1797, in *The Church Teaches*, trans. John F. Clarkson, John H. Edwards, William J. Kelly, and John J. Welch (St. Louis: B. Herder Book Co., 1955), p. 33.

2. *Ibid.*

3. Ernst Mayr, *Animal Species and Evolution* (Cambridge, Mass. and London, England: The Belknap Press of Harvard University Press, 1963), pp. 622-662; T. A. Goudge, *The Ascent of Life* (Toronto: University of Toronto Press, 1961), pp. 133-151; Kenneth F. Weaver, "The Search for Our Ancestors," in *National Geographic*, 168:5 (November 1985), pp. 560-623; C. Loring Brace, "Humans in Time and Space," in *Scientists Confront Creationism*, ed. Laurie R. Godfrey (New York and London: W. W. Norton & Company, 1983), pp. 245-282.

4. Mayr, *Animal Species and Evolution*, p. 635.

5. *Ibid.*, p. 633.

6. *Ibid.*

7. Weaver, "The Search for Our Ancestors," p. 595.

8. Goudge, *The Ascent of Life*, p. 141.

9. Brace, "Humans in Time and Space," in *Scientists Confront Creationism*, p. 260.

10. *Ibid.*

11. *Ibid.*, p. 255.

12. *Ibid.*, p. 261.

13. *Ibid.*, p. 260.

14. *Ibid.*, p. 258.

15. *Ibid.*, p. 264.

16. *Ibid.*, p. 261.

17. Weaver, "The Search for Our Ancestors," p. 609.

18. Brace, "Humans in Time and Space," in *Scientists Confront Creationism*, p. 266.

19. Weaver, "The Search for Our Ancestors," p. 609.

20. *Ibid*.

21. Mayr, *Animal Species and Evolution*, p. 635.

Chapter Ten

1. *Acta apostolis sedis*, 1 (1909), pp. 567-569, as translated in *Rome and the Study of Scripture*, 7th edition (St. Meinrad, Ind.: Abbey Press Publishing Division, 1964), p. 123.

2. *Ibid*.

3. Henricus Denzinger, *Enchiridion symbolorum*, 29th edition, 428, in *The Church Teaches*, trans. John F. Clarkson, John H. Edwards, William J. Kelly, and John J. Welch (St. Louis: B. Herder Book Co., 1955), p. 146.

4. Ludwig Ott, *Fundamentals of Catholic Dogma*, 6th edition (St. Louis: B. Herder Book Co. , 1964), p. 94.

5. Denzinger, *Enchiridion symbolorum*, 2327, in *The Church Teaches*, p. 154.

6. Cyril Vollert, *Symposium on Evolution* (Pittsburgh and Louvain: Duquesne University, Editions E. Nauwelaerts, 1959), p. 102.

7. *Ibid*., p. 98.

8. Ott, *Fundamentals of Catholic Dogma*, p. 95.

9. Denzinger, *Enchiridion symbolorum*, 2328, in *The Church Teaches*, p. 155.

10. Ott, *Fundamentals of Catholic Dogma*, p. 96.

11. Denzinger, *Enchiridion symbolorum*, 2328, in *The Church Teaches*, p. 155.

12. *Ibid*., 790, in *The Church Teaches*, p. 159.

13. Vollert, *Symposium on Evolution*, p. 110.

14. *Ibid*.

15. *Ibid*.

16. St. Thomas Aquinas, *Summa theologiae*, 1, q. 95, a. 1, c (Ottawa: Collège Dominicain d'Ottawa, 1941-1945).

17. Ott, *Fundamentals of Catholic Dogma*, p. 104.

18. *Ibid.*

19. *Ibid.*

20. Pierre Teilhard de Chardin, *The Phenomenon of Man*, trans. Bernard Wall (New York: Harper & Brothers Publishers, 1959), p. 185, note.

21. Raymond J. Nogar, *The Wisdom of Evolution* (Garden City, N.Y.: Doubleday & Company, Inc., 1963), p. 381.

22. *Acta apostolis sedis*, 1 (1909), pp. 567-569, as translated in *Rome and the Study of Scripture*, p. 123.

23. *Ibid.*, p. 124.

24. *Acta apostolis sedis*, 35 (1943), pp. 297-326, as translated in *Rome and the Study of Scripture*, pp. 101-102.

25. St. Thomas Aquinas, *Summa theologiae*, 1, q. 95, a. 1, ad 4.

26. *Ibid.*, a. 3, c.

27. Vollert, *Symposium on Evolution*, p. 114.

28. Nogar, *The Wisdom of Evolution*, p. 382.

29. John A. Hardon, *The Catholic Catechism* (Garden City, N.Y.: Doubleday & Company, Inc., 1975), p. 92.

30. *Ibid.*

31. *Ibid.*

32. *Ibid.*

33. *Ibid.*, p. 93.

34. *Ibid.*

35. *Ibid.*

36. *Ibid.*, p. 98.

37. *Ibid.*, p. 99.

Chapter Eleven

1. C. Loring Brace, "Humans in Time and Space," in *Scientists Confront Creationism*, ed. Laurie R. Godfrey (New York and London: W. W. Norton & Company, 1983), p. 270.

2. Kenneth F. Weaver, "The Search for Our Ancestors," in *National Geographic*, 168:5 (November 1985), p. 598.

3. Jane Goodall, *The Chimpanzees of Gombe* (Cambridge, Mass. and London, England: The Belknap Press of Harvard University Press, 1986), p. 535.

4. *Ibid.*, p. 536.

5. *Ibid.*

6. *Ibid.*, p. 535.

7. *Ibid.*, p. 544.

8. Ernst Mayr, *Animal Species and Evolution* (Cambridge, Mass. and London, England: The Belknap Press of Harvard University Press, 1963), p. 634.

9. Goodall, *The Chimpanzees of Gombe*, p. 535.

10. Brace, "Humans in Time and Space," in *Scientists Confront Creationism*, p. 260.

11. Weaver, "The Search for Our Ancestors," p. 605.

12. Charles E. Oxnard, "Human Fossils: New Views of Old Bones," in *The American Biology Teacher*, 41:5 (May 1979), p. 274.

13. Charles E. Oxnard, "The Place of the Australopithecines in Human Evolution: Grounds for Doubt?", in *Nature*, 258 (4 December 1975), pp. 389-395.

14. *Ibid.*

15. Brace, "Humans in Time and Space," in *Scientists Confront Creationism*, p. 266.

16. *Ibid.*, p. 263.

17. *Ibid.*, p. 264.

18. Kathy Schick and Nick Toth, "Fire," in *Encyclopedia of Human Evolution and Prehistory*, ed. Ian Tattersall, Eric Delson, and John Van Couvering (New York & London: Garland Publishing, 1988), p. 208.

19. Weaver, "The Search for Our Ancestors," p. 609.

20. Schick and Toth, "Fire," in *Encyclopedia of Human Evolution and Prehistory*, p. 208.

21. *Ibid.*

22. John E. Pfeiffer, *The Emergence of Man* (New York, Evanston, and London: Harper & Row, Publishers, 1969), p. 105.

23. *Ibid.*

24. Richard Leakey and Alan Walker, "*Homo Erectus* Unearthed," in *National Geographic*, 168:5 (November 1985), p. 629.

Chapter Twelve

1. Cyril Vollert, *Symposium on Evolution* (Pittsburgh and Louvain: Duquesne University, Editions E. Nauwelaerts, 1959), p. 106.

2. *Ibid.*

3. Raymond J. Nogar, *The Wisdom of Evolution* (Garden City, N.Y.: Doubleday & Company, Inc., 1963), p. 379.

4. *Ibid.*, p. 383.

5. *Ibid.*

6. *Ibid.*

7. *Ibid.*, pp. 383-384.

8. Vollert, *Symposium on Evolution*, pp. 114-115.

9. Pierre Teilhard de Chardin, *The Phenomenon of Man*, trans. Bernard Wall (New York: Harper & Brothers Publishers, 1959), p. 185, note.

10. St. Thomas Aquinas, *Summa theologiae*, 1, q. 91, a. 2, ad. 1 and q. 90, a. 2, c (Ottawa: Collège Dominicain d'Ottawa, 1941-1945).

11. Vollert, *Symposium on Evolution*, p. 109.

12. *Ibid.*, p. 108.

13. *Ibid.*

14. *Ibid.*

15. *Ibid.*

16. *Ibid.*, pp. 88-90.

17. *Ibid.*, p. 99.

18. Nogar, *The Wisdom of Evolution*, p. 380. See, also, Robert T. Francoeur, *Perspectives in Evolution* (Baltimore, Md. and Dublin: Helicon Press, Inc., 1965), pp. 182-183.

19. Peter Damian Fehlner, "In the Beginning...Part III," in *Christ to the World* (The Sacred Congregation for the Evangelization of Peoples) as cited in *Apropos*, 5, Supplement, p. 37, note 5.

20. Vollert, *Symposium on Evolution*, p. 98.

21. Francoeur, *Perspectives in Evolution*, pp. 182-183.

22. *Webster's Seventh New Collegiate Dictionary* (Springfield, Mass.: G. & C. Merriam Company, 1965), p. 270.

23. Henricus Denzinger, *Enchiridion symbolorum*, 29th edition, 6, 86, 255, 256, 429, 462, 993, in *The Church Teaches*, trans. John F. Clarkson, John H. Edwards, William J. Kelly, and John J. Welch (St. Louis: B. Herder Book Co., 1955), pp. 1, 2, 181, 190-191, 205, 206.

24. Rebecca L. Cann, Mark Stoneking, and Allan C. Wilson, "Mitochondrial DNA and Human Evolution," *Nature*, 325 (1 January 1987), p. 31.

25. *Ibid.*, p. 31.

26. *Ibid.*, p. 33.

27. *Ibid.*

28. *Ibid.*, p. 34.

29. *Ibid.*, pp. 35-36.

30. Michael H. Brown, *The Search for Eve* (New York: Harper & Row, Publishers, 1990).

31. *Ibid.*, p. 110.

32. *Ibid.*, p. 108.

33. *Ibid.*

34. *Ibid.*

35. *Ibid.*, p. 270.

36. *Ibid.*, pp. 270-274.

Chapter Thirteen

1. William Henry Green, "Primeval Chronology," in Robert C. Newman and Herman J. Eckelmann, Jr., *Genesis One & the Origin of the Earth* (Downers Grove, Ill.: InterVarsity Press, 1977), pp. 105-123.

2. *Ibid.*, pp. 106-108, 111-114.

3. *Ibid.*, pp. 108-111.

4. *Ibid.*, pp. 115-116.

5. *Ibid.*, pp. 116-118.

6. *Ibid.*, pp. 119-120.

7. *Ibid.*, p. 122.

8. *Ibid.*, p. 123.

9. J. W. G. Johnson, *The Crumbling Theory of Evolution* (Los Angeles, Cal.: Perpetual Eucharistic Adoration, Inc., 1986), p. 71.

10. R. Monastersky, "Deep Desires in Antarctica and Greenland," in *Science News*, 149:22 (1 June 1996), p. 341.

11. Daniel E. Wonderly, "Nonradiometric Data Relevant to the Question of Age," in *Genesis One & the Origin of the Earth*, pp. 89-102.

12. Walter T. Brown, Jr., *In the Beginning* (Phoenix, Ariz.: The Center for Scientific Creation, 1989, p. 19.

13. John Boslough, "The Enigma of TIME," in *National Geographic*, 177:3 (March 1990), p. 127.

14. Trevor Norman and Barry Setterfield, *The Atomic Constants, Light, and Time* (Blackwood, South Australia: By the Authors, 1987), p. 3.

15. *Ibid.*, p. 27.

16. *Ibid.*, pp. 51-55.

17. Johnson, *The Crumbling Theory of Evolution*, pp. 112-113.

18. Ruth Cranston, *The Miracle of Lourdes* (New York: Popular Library, 1957), p. 131.

19. *Ibid.*, p. 133.

20. Johnson, *The Crumbling Theory of Evolution*, p. 113.

21. *Ibid.*

22. W. T. Brown, *In the Beginning*, p. 91.

23. Johnson, *The Crumbling Theory of Evolution*, p. 113.

Chapter Fourteen

1. J. W. G. Johnson, *The Crumbling Theory of Evolution* (Los Angeles, Cal.: Perpetual Eucharistic Adoration, Inc., 1986), p. 55.

2. Walter T. Brown, Jr., *In the Beginning* (Phoenix, Ariz.: The Center for Scientific Creation, 1989), pp. 6, 39-40.

3. Michael A. Cremo and Richard L. Thompson, *Forbidden Archeology* (Los Angeles, Sydney, Stockholm, Bombay: Bhaktivedanta Book Publishing, Inc., 1996).

4. Michael A. Cremo, *Forbidden Archeology's Impact* (Los Angeles, Sydney, Stockholm, Bombay: Bhaktivedanta Book Publishing, Inc., 1998), pp. 344, 349.

5. Jo Wodak and David Oldroyd, "'Vedic Creationism': A Further Twist to the Evolution Debate," in *Social Studies of Science*, 26:1 (1996), p. 207.

6. *Ibid.*

7. *Ibid.*

8 Cremo and Thompson, *Forbidden Archeology*, pp. xxviii, xxx xxxii.

9. *Ibid.*, p. 150.

10. *Ibid.*

11. *Ibid.*

12. *Ibid.*

13. *Ibid.*, pp. 362-366.

14. *Ibid.*, pp. 14-15.

15. *Ibid.*, pp. 19-21.

16. *Ibid.*, p. 630.

17. *Ibid.*, p. 633.

18. *Ibid.*, p. 630-631.

19. *Ibid.*, p. 633.

20. *Ibid.*, p. 631.

21. *Ibid.*, pp. 632-634.

22. *Ibid.*, p. 639.

23. *Ibid.*, p. 793.

24. *Ibid.*, p. 422.

25. *Ibid.*, p. 427.

26. *Ibid.*

27. *Ibid.*, pp. 747.

28. *Ibid.*

29. *Ibid.*, pp. 336-337.

30. *Ibid.*, pp. 194.

31. *Ibid.*, p. 338.

32. *Ibid.*, pp. 368-393.

33. *Ibid.*, p. 368.

34. *Ibid.*, p. 370.

35. *Ibid.*, p. 390.

36. *Ibid.*, pp. 354-366.

37. *Ibid.*, pp. 461-500.

38. *Ibid.*, pp. 555-557.

39. *Ibid.*, pp. 557-558.

40. *Ibid.*, p. 375.

41. *Ibid.*, p. 410.

42. *Ibid.*, p. 445.

43. *Ibid.*, p. 690.

Epilogue

1. Bernard Ramm, *The Christian View of Science and Scripture* (Grand Rapids, Mich.: William B. Eerdmans Publishing Company, 1954), p. 78.

2. Joan Carroll Cruz, *The Incorruptibles* (Rockford, Ill.: Tan Books and Publishers, Inc., 1977).

3. *Ibid.*, p. 27.

4. *Ibid.*

5. *Ibid.*, pp. 288-290.

6. G. K. Chesterton, *George Bernard Shaw* (New York: John Lane Company, 1910), p. 183.

BIBLIOGRAPHY

Aquinas, St. Thomas. *An Introduction to the Philosophy of Nature*. Trans. Roman A. Kocourek. St. Paul, Minn.: North Central Publishing Co., 1956.

_____. *On Being and Essence*. Trans. Armand Maurer. Toronto: The Pontifical Institute of Mediaeval Studies, 1949.

_____. *Quaestiones disputatae de potentia dei*, in St. Thomas Aquinas's *Quaestiones disputatae*. Vol. 22. Ed. P. Bazzi *et al.* 10th edition. Turin-Rome: Marietti, 1965.

_____. *Quaestiones disputatae de veritate*. Vol. 1. Ed. R. Spiazzi. Turin-Rome: Marietti, 1964.

_____. *Summa contra gentiles*. Vols. 13-15, in *Sancti Thomae Aquinatis doctoris angelici opera omnia iussu Leonis XIII. O.M. edita, cura et studio fratrum praediatorum*, Rome, 1918.

_____. *Summa theologiae*. 5 vols. Ottawa: Collège Dominicain d'Ottawa, 1941-1945.

Aristotle. *The Basic Works of Aristotle*. Ed. Richard McKeon. New York: Random House, 1941.

Baum, R. F. "Coming to Grips with Darwin," in *The Intercollegiate Review*, 11:1 (Fall 1975).

Behe, Michael J. *Darwin's Black Box*. New York: The Free Press, 1996.

Bobik, Joseph. *Aquinas on Being and Essence*. Notre Dame, Ind.: University of Notre Dame Press, 1965.

Bonnette, Dennis. *Aquinas' Proofs for God's Existence*. The Hague: Martinus-Nijhoff, 1972.

_____. "How 'Creation' Implies God," in *Faith & Reason*, 11:3-4 (1985).

_____. "Hylomorphism, Positivism, and the Question of When Human Life Begins" in *Social Justice Review*, 72:3-4 (March-April 1981).

_____. "A Philosophical Critical Analysis of Recent Ape-Language Studies," in *Faith & Reason*, 19:2, 3 (Fall 1993).

_____. "A Variation on the First Way of St. Thomas Aquinas," in *Faith & Reason*, 8:1 (Spring 1982).

Boslough, John. "The Enigma of TIME," in *National Geographic*, 177:3 (March 1990).

Bouissac, Paul. "Behavior in Context: In What Sense Is a Circus Animal Performing?", in *The Clever Hans Phenomenon: Communication with Horses, Whales, Apes, and People*. Eds. Thomas A. Sebeok and Robert Rosenthal. Annals of the New York Academy of Sciences. Vol. 364. New York: The New York Academy of Sciences, 1981.

Brace, C. Loring. "Humans in Time and Space," in *Scientists Confront Creationism*. Ed. Laurie R. Godfrey. New York and London: W. W. Norton & Company, 1983.

Brown, Michael H. *The Search for Eve*. New York: Harper & Row, Publishers, 1990.

Brown, Walter T., Jr. *In the Beginning*. Phoenix, Ariz.: Center for Scientific Creation, 1989.

Cann, Rebecca L., Mark Stoneking, and Allan C. Wilson. "Mitochondrial DNA and Human Evolution" in *Nature*, 325 (1 January 1987).

Chesterton, G. K. *The Everlasting Man*. New York: Doubleday & Company, Inc., 1955.

_____. *George Bernard Shaw*. New York: John Lane Company, 1910.

Chevalier-Skolnikoff, Suzanne. "The Clever Hans Phenomenon, Cuing, and Ape Signing: A Piagetian Analysis of Methods for Instructing Animals," in *The Clever Hans Phenomenon: Communication with Horses, Whales, Apes, and People*. Eds. Thomas A. Sebeok and Robert Rosenthal. Annals of the New York Academy of Sciences. Vol. 364. New York: The New York Academy of Sciences, 1981.

Collin, Rémy. *L'Evolution: Hypothèses et Problèmes*. Paris: Librairie Arthème Fayard, 1958.

Cracraft, Joel L. "Systematics, Comparative Biology, and Creationism," in *Scientists Confront Creationism*. Ed. Laurie R. Godfrey. New York and London: W. W. Norton & Company, 1983.

Cranston, Ruth. *The Miracle of Lourdes*. New York: Popular Library, 1957.

Cremo, Michael A. and Richard L. Thompson. *Forbidden Archeology*. Los Angeles, Sydney, Stockholm, Bombay: Bhaktivedanta Book Publishing, Inc., 1996.

Cremo, Michael A. *Forbidden Archeology's Impact*. Los Angeles, Sydney, Stockholm, Bombay: Bhaktivedanta Book Publishing, Inc., 1998.

Crowley, Charles B. *Aristotelian-Thomistic Philosophy of Measure and the International System of Units (SI)*. Ed. with a prescript by Peter A. Redpath. Lanham, Md.: University Press of America, 1996.

Cruz, Joan Carroll. *The Incorruptibles*. Rockford, Ill.: Tan Books and Publishers, Inc., 1977.

Cuffey, Roger J. "Paleontologic Evidence and Organic Evolution," in *Science and Creationism*. Ed. Ashley Montagu. Oxford, New York, Toronto, Melbourne: Oxford University Press, 1984.

Darwin, Charles. *On the Origin of Species*. Cambridge, Mass. and London, England: Harvard University Press, 1964.

Dawkins, Richard. *The Blind Watchmaker*. New York: W. W. Norton & Company, 1996.

Deely, John N. *The Philosophical Dimensions of the Origin of Species*. Chicago: Institute for Philosophical Research, 1969.

Denton, Michael. *Evolution: A Theory in Crisis*. Bethesda, Md.: Adler & Adler, Publishers, Inc., 1985.

Denzinger, Henricus. *Enchiridion symbolorum*, in *The Church Teaches*. Trans. John F. Clarkson, John H. Edwards, William J. Kelly, and John J. Welch. St. Louis: B. Herder Book Co., 1955.

Dieska, Joseph. "The Intuitivism of N. O. Lossky," in *University of Dayton Review*, 7:2 (Spring 1971).

Doolittle, Russell F. "Probability and the Origin of Life," in *Scientists Confront Creationism*. Ed. Laurie R. Godfrey. New York and London: W. W. Norton & Company, 1983.

Eddington, Sir Arthur. *The Nature of the Physical World*. New York: The Macmillan Company and Cambridge, England: The University Press, 1929.

Encrenaz, T. and Bibring, J.-P. *The Solar System*. Berlin, Heidelberg, New York, London, Paris, Tokyo, Hong Kong: Springer-Verlag, 1990.

Field, A. N. *The Evolution Hoax Exposed*. Rockford, Ill.: Tan Books and Publishers, 1971.

Flew, Anthony. *God: A Critical Enquiry*. La Salle, Ill.: Open Court Publishing Company, 1984.

Fox, Sidney W. "Creationism and Evolutionary Protobiogenesis," in *Science and Creationism*. Ed. Ashley Montagu. Oxford, New York, Toronto, Melbourne: Oxford University Press, 1984.

Francoeur, Robert T. *Perspectives in Evolution*. Baltimore, Md. and Dublin: Helicon Press, Inc., 1965.

Gallant, Roy A. "To Hell with Evolution," in *Science and Creationism*. Ed. Ashley Montagu. Oxford, New York, Toronto, Melbourne: Oxford University Press, 1984.

Garrigou-Lagrange, Reginald. *God: His Existence and His Nature*. Trans. Dom Bede Rose. St. Louis: B. Herder Book Co., 1934.

_____. *The One God*. Trans. Dom Bede Rose. St. Louis: B. Herder Book Co., 1959.

Gerrity, Brother Benignus. *Nature, Knowledge, and God*. Milwaukee: The Bruce Publishing Company, 1947.

Gilson, Etienne. *Being and Some Philosophers*. Toronto: The Pontifical Institute of Mediaeval Studies, 1952.

_____. *The Christian Philosophy of St. Thomas Aquinas*. Trans. L. K. Shook. New York: Random House, 1956.

_____. *The Elements of Christian Philosophy*. New York-Toronto: The New American Library, 1960.

_____. *The Philosophy of St. Thomas Aquinas (Le Thomisme)*. Trans. Edward Bullough. Cambridge, England: W. Heffer & Sons, Ltd., 1929.

_____. *The Spirit of Mediaeval Philosophy*. Trans. A. H. C. Downes. New York: Charles Scribner's Sons, 1940.

Godfrey, Laurie R. (ed.) *Scientists Confront Creationism*. New York and London: W. W. Norton & Company, 1983.

Goodall, Jane. *The Chimpanzees of Gombe*. Cambridge, Mass. and London, England: The Belknap Press of Harvard University Press, 1986.

Goudge, T. A. *The Ascent of Life*. Toronto: University of Toronto Press, 1961.

Gould, Stephen Jay. *Hen's Teeth and Horse's Toes*. New York-London: W. W. Norton & Company, 1983.

Graham-Smith, Sir Francis and Sir Bernard Lovell. *Pathways to the Universe*. Cambridge, New York, New Rochelle, Melbourne, Sydney: Cambridge University Press, 1988.

Green, William Henry. "Primeval Chronology," in Robert C. Newman and Herman J. Eckelmann, Jr. *Genesis One & the Origin of the Earth*. Downers Grove, Ill.: InterVarsity Press, 1977.

Hardon, John A. *The Catholic Catechism*. Garden City, N.Y.: Doubleday & Company, Inc., 1975.

Hediger, Heini K. P. "The Clever Hans Phenomenon from an Animal Psychologist's Point of View," in *The Clever Hans Phenomenon: Communication with Horses, Whales, Apes, and People*. Eds. Thomas A. Sebeok and Robert Rosenthal. Annals of the New York Academy of Sciences. Vol. 364. New York: The New York Academy of Sciences, 1981.

The Holy Bible. Douay-Rheims. Revised by Richard Challoner. Rockford, Ill.: Tan Books and Publishers, Inc., 1989.

Hume, David. *A Treatise of Human Nature*. London: J. M. Dent & Sons, Ltd., 1956.

Huxley, Julian. *Evolution in Action*. New York: New American Library, 1953.

Jaki, Stanley L. *Brain, Mind, and Computers*. New York: Herder and Herder, Inc., 1969.

Jastrow, Robert. *God and the Astronomers*. New York: W. W. Norton and Co., Inc., 1978.

Johanson, Donald C., *et al*. "New Partial Skeleton of *homo habilis* from Olduvai Gorge, Tanzania" in *Nature*, 327 (21 May 1987).

Johnson, J. W. G. *The Crumbling Theory of Evolution*. Los Angeles: Perpetual Eucharistic Adoration, Inc., 1986.

Johnson, Phillip E. *Darwin on Trial*. Washington: Regnery Gateway, 1991.

Kramer, William. *Evolution & Creation: A Catholic Understanding*. Huntington, Ind.: Our Sunday Visitor Publishing Division, Our Sunday Visitor, Inc., 1986.

Leakey, Richard. "Further Evidence of Lower Pleistocene Hominids from East Rudolf, North Kenya," in *Nature*, 231 (28 May 1971).

Leakey, Richard and Alan Walker. "Homo Erectus Unearthed" in *National Geographic*, 168:5 (November 1985).

Lecomte du Noüy, Pierre. *The Road to Reason*. New York and Toronto: Longmans, Green, and Co., 1948.

Lovell, Sir Bernard. *In the Center of Immensities*. New York, Hagerstown, San Francisco, London: Harper & Row, Publishers, 1978.

Maritain, Jacques. *Approaches to God*. New York: Collier Books, 1962.

_____. *The Degrees of Knowledge*. Trans. Gerald B. Phelan. New York: Charles Scribner's Sons, 1959.

_____. *A Preface to Metaphysics*. New York: Sheed & Ward, 1939.

Mayr, Ernst. *Animal Species and Evolution*. Cambridge, Mass. and London: The Belknap Press of Harvard University Press, 1963.

_____. *The Species Problem*. Washington: American Association for the Advancement of Science, 1957.

McMullin, Ernan. (ed.) *Evolution and Creation*. Notre Dame, Ind.: University of Notre Dame Press, 1985.

Monastersky, R. "Deep Desires in Antarctica and Greenland," in *Science News*, 149:22 (1 June 1996).

Montagu, Ashley. (ed.) *Science and Creationism*. Oxford, New York, Toronto, Melbourne: Oxford University Press, 1984.

Nahm, Milton C. *Selections from Early Greek Philosophy*. New York: Appleton-Century-Crofts, 1964.

Newman, Robert C. and Herman J. Eckelmann, Jr. *Genesis One & the Origin of the Earth*. Downers Grove, Ill.: InterVarsity Press, 1977.

Nogar, Raymond J. *The Wisdom of Evolution*. Garden City, N.Y.: Doubleday & Co., Inc., 1963.

Norman, Trevor and Barry Setterfield. *The Atomic Constants, Light, and Time*. Blackwood, South Australia: By the Authors, 1987.

Nourse, Alan E. *Universe, Earth, and Atom*. New York and Evanston: Harper & Row, Publishers, 1969.

Ott, Ludwig. *Fundamentals of Catholic Dogma*. 6th edition. St. Louis: B. Herder Book Co., 1964.

Oxnard, Charles E. "Human Fossils: New Views of Old Bones," in *The American Biology Teacher*, 41:5 (May 1979).

_____. "The Place of the Australopithecines in Human Evolution: Grounds for Doubt?", in *Nature*, 258 (4 December 1975).

Patterson, John W. "Thermodynamics and Evolution," in *Scientists Confront Creationism*. Ed. Laurie R. Godfrey. New York and London: W. W. Norton & Company, 1983.

Pegis, Anton C. (ed.). *Basic Writings of Saint Thomas Aquinas*. Trans. Laurence Shapcote. 2 vols. New York: Random House, 1945.

Pfeiffer, John E. *The Emergence of Man*. New York, Evanston, and London: Harper & Row, Publishers, 1969.

Ramm, Bernard. *The Christian View of Science and Scripture*. Grand Rapids, Mich.: William B. Eerdmans Publishing Company, 1954.

Ringwood, A. E. *Origin of the Earth and Moon*. New York, Heidelberg, Berlin: Springer-Verlag, 1979.

Rome and the Study of Scripture. 7th edition. St. Meinrad, Ind.: Abbey Press Publishing Division, 1964.

Ross, Hugh. *The Creator and the Cosmos*. Colorado Springs, Col.: NavPress, 1995.

Rumbaugh, Duane M. "Who Feeds Clever Hans?", in *The Clever Hans Phenomenon: Communication with Horses, Whales, Apes, and People*. Eds. Thomas A. Sebeok and Robert Rosenthal. Annals of the New York Academy of Sciences. Vol. 364. New York: The New York Academy of Sciences, 1981.

Savage-Rumbaugh, Sue E. "Can Apes Use Symbols to Represent Their World?", in *The Clever Hans Phenomenon: Communication with Horses, Whales, Apes, and People*. Eds. Thomas A. Sebeok and Robert Rosenthal. Annals of the New York Academy of Sciences. Vol. 364. New York: The New York Academy of Sciences, 1981.

Sebeok, Thomas A. and Jean Umiker-Sebeok. "Performing Animals: Secrets of the Trade," in *Psychology Today*, 13:6 (November 1979).

Sebeok, Thomas A. and Robert Rosenthal. (eds.) *The Clever Hans Phenomenon: Communication with Horses, Whales, Apes, and People*. Annals of the New York Academy of Sciences. Vol. 364. New York: The New York Academy of Sciences, 1981.

Seidenberg, Mark S. and Laura A. Petitto. "Ape Signing: Problems of Method and Interpretation," in *The Clever Hans Phenomenon: Communication with Horses, Whales, Apes, and People*. Eds. Thomas A. Sebeok and Robert Rosenthal. Annals of the New York Academy of Sciences. Vol. 364. New York: The New York Academy of Sciences, 1981.

Shores, Louis (ed.). *Collier's Encyclopedia*. Ed. Louis Shores. Vol. 9. Canada: P. F. Collier's & Son, Ltd., 1973.

Sillem, Edward. *Ways of Thinking about God*. New York: Sheed and Ward, 1961.

Tattersall, Ian, Eric Delson, and John Van Couvering (eds). *Encyclopedia of Human Evolution and Prehistory*. New York & London: Garland Publishing, 1988.

Taylor, R. E., *et al*. "Major Revisions in the Pleistocene Age Assignments for North American Human Skeletons by C-14 Accelerator Mass Spectrometry: None Older than 11,000 C-14 Years B.P.," in *American Antiquity*, 50:1 (1985).

Tax, Sol. (ed.) *Evolution after Darwin*. Chicago: University of Chicago Press, 1960.

Teilhard de Chardin, Pierre. *The Phenomenon of Man*. Trans. Bernard Wall. New York: Harper & Brothers Publishers, 1959.

Terrace, Herbert S. "How Nim Chimpsky Changed My Mind," in *Psychology Today*, 13:6 (November 1979).

_____. "In the Beginning Was the 'Name,'" in *American Psychologist*, 40:9 (September 1985).

Troitskii, V. S. "Physical Constants and Evolution of the Universe," in *Astrophysics and Space Science*, 139 (August 1987).

Van Flandern, T. C. "Is the Gravitational Constant Changing?", *The Astrophysical Journal*, 248 (1 September 1981).

Velikovsky, Immanuel. *Earth in Upheaval*. New York: Dell Publishing Co., Inc., 1955.

Vollert, Cyril. *Symposium on Evolution*. Pittsburgh and Louvain: Duquesne University-Editions E. Nauwelaerts, 1959.

Walker, Stephen. *Animal Thought*. London, Boston, Melbourne, Henley: Routledge & Kegan Paul, Ltd., 1983.

Weaver, Kenneth F. "The Search for Our Ancestors," in *National Geographic*, 168:5 (November 1985).

Webster's Seventh New Collegiate Dictionary. Ed. in Chief Philip B. Gove. Springfield, Mass.: G. & C. Merriam Company, 1965.

Wilders, Peter. "Creation *Ex Nihilo*," in *The Wanderer* (6 May 1999).

Wodak, Jo and David Oldroyd. "'Vedic Creationism': A Further Twist to the Evolution Debate," in *Social Studies of Science*, 26:1 (1996).

Wonderly, Daniel E. "Nonradiometric Data Relevant to the Question of Age," in Robert C. Newman and Herman J. Eckelmann, Jr. *Genesis One & the Origin of the Earth*. Downers Grove, Ill.: InterVarsity Press, 1977.

Woodbury, Austin M. *Cosmology*. Sydney, N.S.W.: Aquinas Academy, unpublished manuscript, 1949.

_____. *Natural Philosophy, Treatise Three, Psychology*. Bk. 3. Sydney, N.S.W.: Aquinas Academy, unpublished manuscript, 1951.

_____. *Philosophical Psychology*. Sydney, N.S.W.: Aquinas Academy, unpublished manuscript, 1945 and 1951.

Wuellner, Bernard. *Dictionary of Scholastic Philosophy*. Milwaukee: The Bruce Publishing Company, 1956.

Zuckerman, Sir Solly. *Beyond the Ivory Tower*. New York: Taplinger Publishing Company, 1970.

ABOUT THE AUTHOR

Dennis Bonnette is a Full Professor of Philosophy and currently Chairman of the Philosophy Department at Niagara University in Lewiston, New York. He lives in Youngstown, New York, with his wife, Lois. He is a former National Defense Education Act Fellow at the University of Notre Dame, from which he received his doctorate in Philosophy in 1970. He has been teaching Philosophy at the college level for 37 years. Dr. Bonnette has written one earlier book, *Aquinas' Proofs for God's Existence,* and many scholarly articles. He is a member of the American Philosophical Association, Scholars for Social Justice, and the American Catholic Philosophical Association. He is especially interested in using philosophical analysis to reconcile enigmas that appear to arise from the interface of natural science and divine revelation.

INDEX

VIBS

The **Value Inquiry Book Series** is co-sponsored by:

American Maritain Association
American Society for Value Inquiry
Association for Process Philosophy of Education
Center for Bioethics, University of Turku
Center for International Partnerships, Rochester Institute of Technology
Center for Professional and Applied Ethics, University of North Carolina at
Charlotte
Centre for Applied Ethics, Hong Kong Baptist University
Centre for Cultural Research, Aarhus University
College of Education and Allied Professions, Bowling Green State University
Concerned Philosophers for Peace
Conference of Philosophical Societies
Gannon University
Global Association for the Study of Persons
Institute of Philosophy of the High Council of Scientific Research, Spain
International Academy of Philosophy of the Principality of Liechtenstein
International Center for the Arts, Humanities, and Value Inquiry
International Society for Universal Dialogue
Natural Law Society
Philosophical Society of Finland
Philosophy Born of Struggle Association
Philosophy Seminar, University of Mainz
Pragmatism Archive
R.S. Hartman Institute for Formal and Applied Axiology
Russian Philosophical Society
Society for Iberian and Latin-American Thought
Society for the Philosophic Study of Genocide and the Holocaust
Society for the Philosophy of Sex and Love
Yves R. Simon Institute.

Titles Published

1. Noel Balzer, *The Human Being as a Logical Thinker.*

2. Archie J. Bahm, *Axiology: The Science of Values.*

3. H. P. P. (Hennie) Lötter, *Justice for an Unjust Society.*

4. H. G. Callaway, *Context for Meaning and Analysis: A Critical Study in the Philosophy of Language.*

5. Benjamin S. Llamzon, *A Humane Case for Moral Intuition.*

6. James R. Watson, *Between Auschwitz and Tradition: Postmodern Reflections on the Task of Thinking.* A volume in **Holocaust and Genocide Studies.**

7. Robert S. Hartman, *Freedom to Live: The Robert Hartman Story,* edited by Arthur R. Ellis. A volume in **Hartman Institute Axiology Studies.**

8. Archie J. Bahm, *Ethics: The Science of Oughtness.*

9. George David Miller, *An Idiosyncratic Ethics; Or, the Lauramachean Ethics.*

10. Joseph P. DeMarco, *A Coherence Theory in Ethics.*

11. Frank G. Forrest, *Valuemetrics^N: The Science of Personal and Professional Ethics.* A volume in **Hartman Institute Axiology Studies.**

12. William Gerber, *The Meaning of Life: Insights of the World's Great Thinkers.*

13. Richard T. Hull, Editor, *A Quarter Century of Value Inquiry: Presidential Addresses of the American Society for Value Inquiry.* A volume in **Histories and Addresses of Philosophical Societies.**

14. William Gerber, *Nuggets of Wisdom from Great Jewish Thinkers: From Biblical Times to the Present.*

15. Sidney Axinn, *The Logic of Hope: Extensions of Kant's View of Religion.*

16. Messay Kebede, *Meaning and Development.*

17. Amihud Gilead, *The Platonic Odyssey: A Philosophical-Literary Inquiry into the* Phaedo.

18. Necip Fikri Alican, *Mill's Principle of Utility: A Defense of John Stuart Mill's Notorious Proof.* A volume in **Universal Justice.**

19. Michael H. Mitias, Editor, *Philosophy and Architecture.*

20. Roger T. Simonds, *Rational Individualism: The Perennial Philosophy of Legal Interpretation.* A volume in **Natural Law Studies.**

21. William Pencak, *The Conflict of Law and Justice in the Icelandic Sagas.*

22. Samuel M. Natale and Brian M. Rothschild, Editors, *Values, Work, Education: The Meanings of Work.*

23. N. Georgopoulos and Michael Heim, Editors, *Being Human in the Ultimate: Studies in the Thought of John M. Anderson.*

24. Robert Wesson and Patricia A. Williams, Editors, *Evolution and Human Values.*

25. Wim J. van der Steen, *Facts, Values, and Methodology: A New Approach to Ethics.*

26. Avi Sagi and Daniel Statman, *Religion and Morality.*

27. Albert William Levi, *The High Road of Humanity: The Seven Ethical Ages of Western Man,* edited by Donald Phillip Verene and Molly Black Verene.

28. Samuel M. Natale and Brian M. Rothschild, Editors, *Work Values: Education, Organization, and Religious Concerns.*

29. Laurence F. Bove and Laura Duhan Kaplan, Editors, *From the Eye of the Storm: Regional Conflicts and the Philosophy of Peace.* A volume in **Philosophy of Peace.**

59. Leena Vilkka, *The Intrinsic Value of Nature.*

60. Palmer Talbutt, Jr., *Rough Dialectics: Sorokin's Philosophy of Value,* with Contributions by Lawrence T. Nichols and Pitirim A. Sorokin.

61. C. L. Sheng, *A Utilitarian General Theory of Value.*

62. George David Miller, *Negotiating Toward Truth: The Extinction of Teachers and Students.* Epilogue by Mark Roelof Eleveld. A volume in **Philosophy of Education.**

63. William Gerber, *Love, Poetry, and Immortality: Luminous Insights of the World's Great Thinkers.*

64. Dane R. Gordon, Editor, *Philosophy in Post-Communist Europe.* A volume in **Post-Communist European Thought.**

65. Dane R. Gordon and Józef Niznik, Editors, *Criticism and Defense of Rationality in Contemporary Philosophy.* A volume in **Post-Communist European Thought.**

66. John R. Shook, *Pragmatism: An Annotated Bibliography, 1898-1940.* With Contributions by E. Paul Colella, Lesley Friedman, Frank X. Ryan, and Ignas K. Skrupskelis.

67. Lansana Keita, *The Human Project and the Temptations of Science.*

68. Michael M. Kazanjian, *Phenomenology and Education: Cosmology, Co-Being, and Core Curriculum.* A volume in **Philosophy of Education.**

69. James W. Vice, *The Reopening of the American Mind: On Skepticism and Constitutionalism.*

70. Sarah Bishop Merrill, *Defining Personhood: Toward the Ethics of Quality in Clinical Care.*

71. Dane R. Gordon, *Philosophy and Vision.*

72. Alan Milchman and Alan Rosenberg, Editors, *Postmodernism and the Holocaust.* A volume in **Holocaust and Genocide Studies.**

73. Peter A. Redpath, *Masquerade of the Dream Walkers: Prophetic Theology from the Cartesians to Hegel.* A volume in **Studies in the History of Western Philosophy.**

74. Malcolm D. Evans, *Whitehead and Philosophy of Education: The Seamless Coat of Learning.* A volume in **Philosophy of Education.**

75. Warren E. Steinkraus, *Taking Religious Claims Seriously: A Philosophy of Religion,* edited by Michael H. Mitias. A volume in **Universal Justice.**

76. Thomas Magnell, Editor, *Values and Education.*

77. Kenneth A. Bryson, *Persons and Immortality.* A volume in **Natural Law Studies.**

78. Steven V. Hicks, *International Law and the Possibility of a Just World Order: An Essay on Hegel's Universalism.* A volume in **Universal Justice.**

79. E. F. Kaelin, *Texts on Texts and Textuality: A Phenomenology of Literary Art,* edited by Ellen J. Burns.

80. Amihud Gilead, *Saving Possibilities: A Study in Philosophical Psychology.* A volume in **Philosophy and Psychology.**

81. André Mineau, *The Making of the Holocaust: Ideology and Ethics in the Systems Perspective.* A volume in **Holocaust and Genocide Studies.**

82. Howard P. Kainz, *Politically Incorrect Dialogues: Topics Not Discussed in Polite Circles.*

83. Veikko Launis, Juhani Pietarinen, and Juha Räikkä, Editors, *Genes and Morality: New Essays.* A volume in **Nordic Value Studies.**

84. Steven Schroeder, *The Metaphysics of Cooperation: The Case of F. D. Maurice.*

85. Caroline Joan ("Kay") S. Picart, *Thomas Mann and Friedrich Nietzsche: Eroticism, Death, Music, and Laughter.* A volume in **Central-European Value Studies.**

86. G. John M. Abbarno, Editor, *The Ethics of Homelessness: Philosophical Perspectives.*

87. James Giles, Editor, *French Existentialism: Consciousness, Ethics, and Relations with Others.* A volume in **Nordic Value Studies.**

88. Deane Curtin and Robert Litke, Editors, *Institutional Violence.* A volume in **Philosophy of Peace.**

89. Yuval Lurie, *Cultural Beings: Reading the Philosophers of* Genesis.

90. Sandra A. Wawrytko, Editor, *The Problem of Evil: An Intercultural Exploration.* A volume in **Philosophy and Psychology.**

91. Gary J. Acquaviva, *Values, Violence, and Our Future.* A volume in **Hartman Institute Axiology Studies.**

92. Michael R. Rhodes, *Coercion: A Nonevaluative Approach.*

93. Jacques Kriel, *Matter, Mind, and Medicine: Transforming the Clinical Method.*

94. Haim Gordon, *Dwelling Poetically: Educational Challenges in Heidegger's Thinking on Poetry.* A volume in **Philosophy of Education.**

95. Ludwig Grünberg, *The Mystery of Values: Studies in Axiology,* edited by Cornelia Grünberg and Laura Grünberg.

96. Gerhold K. Becker, Editor, *The Moral Status of Persons: Perspectives on Bioethics.* A volume in **Studies in Applied Ethics.**

97. Roxanne Claire Farrar, *Sartrean Dialectics: A Method for Critical Discourse on Aesthetic Experience.*

98. Ugo Spirito, *Memoirs of the Twentieth Century.* Translated from Italian and edited by Anthony G. Costantini. A volume in **Values in Italian Philosophy.**

99. Steven Schroeder, *Between Freedom and Necessity: An Essay on the Place of Value.*

100. Foster N. Walker, *Enjoyment and the Activity of Mind: Dialogues on Whitehead and Education.* A volume in **Philosophy of Education.**

101. Avi Sagi, *Kierkegaard, Religion, and Existence: The Voyage of the Self.* Translated from Hebrew by Batya Stein.

Printed in the United States
by Baker & Taylor Publisher Services

Printed in the United States
by Baker & Taylor Publisher Services